LONG-RANGE CORRELATIONS IN ASTROPHYSICAL SYSTEMS

ANNALS OF THE NEW YORK ACADEMY OF SCIENCES
Volume 848

LONG-RANGE CORRELATIONS IN ASTROPHYSICAL SYSTEMS

Edited by J. Robert Buchler, James W. Dufty, and Henry E. Kandrup

The New York Academy of Sciences
New York, New York
1998

Library of Congress Cataloging-in-Publication Data

Long-range correlations in astrophysical systems / edited by J. Robert Buchler, James W. Dufty, and Henry E. Kandrup.
p. cm. — (Annals of the New York Academy of Sciences ; v. 848).
Includes bibliographical references and index.
ISBN 1-57331-112-X (alk. paper). — ISBN 1-57331-113-8 (pbk. : alk. paper)
1. Astrophysics—Methodology—Congresses. 2. Correlation (Statistics)—Congresses. I. Buchler, J. R. (J. Robert) II. Dufty, James W. III. Kandrup, Henry E. IV. Series.
Q11.N5 vol. 848
[QB460]
500 s—dc21
[523.01] 98-13137
CIP

GYAT / M-V
Printed in the United States of America
ISBN 1-57331-112-X (cloth)
ISBN 1-57331-113-8 (paper)
ISSN 0077-8923

ANNALS OF THE NEW YORK ACADEMY OF SCIENCES

Volume 848
May 15, 1998

LONG-RANGE CORRELATIONS IN ASTROPHYSICAL SYSTEMS[a]

Editors and Conference Organizers
J. ROBERT BUCHLER, JAMES W. DUFTY, AND
HENRY E. KANDRUP

CONTENTS

[a]This volume is the result of the twelfth conference of a series entitled *Florida Workshops in Nonlinear Astronomy* held on February 20–22, 1997 in Gainesville, Florida.

Financial assistance was received from:

- COLLEGE OF LIBERAL ARTS AND SCIENCES AT THE UNIVERSITY OF FLORIDA
- DEPARTMENTS OF PHYSICS AND ASTRONOMY AT THE UNIVERSITY OF FLORIDA
- NATIONAL SCIENCE FOUNDATION
- OFFICE OF RESEARCH, TECHNOLOGY, AND GRADUATE EDUCATION AT THE UNIVERSITY OF FLORIDA

Preface

Cross disciplinary research in physics has become increasingly intense and productive over the past decade. It is clear from many examples that developments in one field can advance more quickly when there is information available regarding parallel developments in other fields. This occurs not only by exchange of methods and techniques but also from conceptual refinement resulting from different perspectives on common problems. One of the most effective means for stimulating such exchange is to assemble expert representatives from several fields for discussion of a loosely defined topic on which there have been recent developments in each field. The 1997 Workshop in Nonlinear Astronomy and Physics held February 20–22 at the University of Florida assembled twenty theorists from the fields of astronomy, mathematics, plasma physics, condensed matter physics, and statistical mechanics for discussions of the general topic *long-range correlations in astrophysical and other systems*. Each of the invited participants presented a lecture indicating the current status of related research from his or her field. Participating students presented posters with an informal oral presentation to the group. The format included ample time for discussion to identify points of overlap and the potential for fruitful exploitation. The primary motivation was to develop cross-talk on large scale structures and related long-range correlations and their associated underlying mechanisms. Examples illustrating the diversity of topics considered include long-range order in gravitational systems, gravitational chaos, correlations in neutral and charged Coulomb systems, long-range correlations in systems with short-range forces in nonequilibrium states, pattern formation due to instability and bifurcation, fluid-type cellular automata without detailed balance, clustering of granular media, fluctuation-induced forces, and generic self-organization.

Support for the speakers and workshop infrastructure was provided by the Office of Research, Technology, and Graduate Education and by the College of Liberal Arts and Sciences at the University of Florida; the National Science Foundation provided additional support for graduate students from six U.S. institutions. We would also like to thank the Editorial Department of the New York Academy of Sciences, and are particularly grateful to Stephanie Bludau for her skill in seeing this volume through the press.

— J. ROBERT BUCHLER
— JAMES W. DUFTY
— HENRY E. KANDRUP

Long-Range Correlations in Nonequilibrium Fluids[a]

JAMES W. DUFTY

Department of Physics, University of Florida, Gainesville, Florida 32611

ABSTRACT: Simple fluids with short-range atomic forces at equilibrium exhibit long-range correlations only at thermodynamic critical points. In contrast, the same fluids have long-range correlations for a wide class of nonequilibrium states. The mechanism behind such nonequilibrium correlations is described here briefly, both from linear response methods and from kinetic theory.

1. INTRODUCTION

The correlation length for spontaneous fluctuations in a simple atomic fluid at equilibrium is typically of the order of the force range. However, near a thermodynamic critical point the correlation length grows without bound as the critical point is approached. The appearance of long-range correlations at special values of thermodynamic parameters has been the subject of intense study over the past twenty years with renormalization group methods being developed and refined to account for the growth of the correlation length. More recently, it has become clear that long-range correlations occur *generically* for nonequilibrium states without the need to "tune" thermodynamic parameters to special critical points. Thus it appears in this more general context that long-range correlations are the norm, with short-range correlations being an exceptional consequence of the special detailed balance conditions for equilibrium states. The objective here is to provide some insight into the origin of long-range correlations in nonequilibrium states and to indicate some methods for their analysis. This is done in two ways. First, the very general methods of linear response for states near equilibrium are applied to study density fluctuations. Second, kinetic theory methods are applied for more general nonequilibrium states of gases at low density.

During the past decade many methods have been applied to analyze long-range correlations in nonequilibrium states. The simplest of these is a stochastic hydrodynamics description. Detailed quantitative analyses have now been compared with experiments for the case of a steady state with a uniform temperature gradient [1]. The agreement is excellent, confirming both the existence of the correlations and the mechanism responsible for them. This mechanism is closely related to "mode coupling" effects that are responsible for anomalous long time decay of fluctuations as well as large spatial correlations, both in the nonequilibrium states and for equilibrium critical points. Further details of this connection are provided in a recent review [2]. The relationship to other mechanisms for long-range order such as avalanches in self-organized criticality is more remote.

[a]This research was supported by NSF grants PHY 9722133 and 9312723.

A convenient property for studying long-range correlations in this discussion is the density autocorrelation function,

$$C(\mathbf{r},\mathbf{r}'; t) \equiv \langle \delta n(\mathbf{r})\delta n(\mathbf{r}'); t\rangle = \int d\Gamma \rho(\Gamma, t)\delta n(\mathbf{r}; \Gamma)\delta n(\mathbf{r}'; \Gamma). \tag{1}$$

Here $\delta n(\mathbf{r}) = n(\mathbf{r}) - \langle n(\mathbf{r})\rangle$ is the deviation of the microscopic local density from its average value. The second equality expresses the fluctuation in terms of an ensemble average over the N particle phase space with $\rho(\Gamma, t)$ representing the nonequilibrium state at time t. The conditions on this state necessary for long-range correlations in nonequilibrium states are described below. First, however, we indicate that some simple possibilities are not sufficient. For example, it might be thought that an equilibrium state with a nonuniform external force could introduce a new length scale for correlations. To see that this is not the case, first express $C(\mathbf{r},\mathbf{r}'; t) \to C_e(\mathbf{r}, \mathbf{r}')$ in terms of the pair correlation function $g(\mathbf{r}, \mathbf{r}')$:

$$C(\mathbf{r},\mathbf{r}'; t) = \langle n(\mathbf{r})\rangle[\delta(\mathbf{r} - \mathbf{r}') + \langle n(\mathbf{r}')\rangle(g(\mathbf{r}, \mathbf{r}') - 1)]. \tag{2}$$

The pair correlation function has a well-known expansion in powers of the density given by

$$g(\mathbf{r}, \mathbf{r}') - 1 \to f(\mathbf{r} - \mathbf{r}') + \int d\mathbf{r}'' f(\mathbf{r} - \mathbf{r}'')\langle n(\mathbf{r}'')\rangle f(\mathbf{r}'' - \mathbf{r}') + \ldots, \tag{3}$$

where $f(\mathbf{r}) \equiv e^{-V(r)/k_BT} - 1$, and $V(r)$ is the short-range interatomic potential. The average density behaves as $\langle n(\mathbf{r})\rangle \sim n_e e^{-V_{ext}(r)/k_BT}$ and has a spatial variation determined by the external force. However, the correlation of two points in the system is determined by $f(\mathbf{r})$ so that the correlation length is still of the order of the interatomic force range. This demonstrates that long-range correlations require a nonequilibrium state, not simply an inhomogeneous equilibrium state. The simplest case is a system driven by inhomogeneous boundary conditions to produce an average current, $\langle \mathbf{j}(\mathbf{r})\rangle$, of mass, energy, or momentum. If the spatial gradient of the current is small then it is proportional to the gradient of a corresponding conjugate thermodynamic field $y(\mathbf{r})$ (chemical potential, temperature, flow velocity), $\langle j(\mathbf{r})\rangle = -\lambda\nabla y(\mathbf{r})$. To calculate $C(\mathbf{r},\mathbf{r}'; t)$ it might be assumed that the ensemble in (1) could be estimated by the corresponding information entropy ensemble, obtained by imposing the constraint that the average current be given by this linear relationship. This prescription is easily carried out and to linear order in the gradient yields

$$C(\mathbf{r},\mathbf{r}'; t) \to C_e(\mathbf{r} - \mathbf{r}') - \lambda\int d\mathbf{r}_1 R(\mathbf{r}, \mathbf{r}', \mathbf{r}_1)_1 \nabla_1 y(\mathbf{r}_1), \tag{4}$$

$$R(\mathbf{r}, \mathbf{r}', \mathbf{r}_1) = \int d\mathbf{r}_2 \langle \delta n(\mathbf{r})\delta n(\mathbf{r}')\mathbf{j}(\mathbf{r}_2)\rangle_e h^{-1}(\mathbf{r}_2, \mathbf{r}_1), \quad h^{-1}(\mathbf{r}_2, \mathbf{r}_1) \equiv \langle \mathbf{j}(\mathbf{r}_1)\mathbf{j}(\mathbf{r}_2)\rangle_e. \tag{5}$$

The subscript e denotes an equilibrium average. It is easily verified that these equilibrium correlation functions all have correlation lengths of the order of the force range, and therefore no long-range effects are possible in this description. The main conclusion from these two examples is that nonequilibrium states are required and that the description of these states must include more than simply their average properties. The following sections indicated that the dynamics of fluctuations must be accounted for at some level.

2. LINEAR RESPONSE

The second example of the last section represents a nonequilibrium state sufficiently close to equilibrium so that deviations in the order parameter, $\nabla y(\mathbf{r})$, can be computed perturbatively. As the estimate using information entropy methods is inadequate we return to the exact result obtained from solution to the Liouville equation [3]. For a one component fluid in steady state the deviations from equilibrium are linear in the temperature and flow velocity gradients [3]

$$\rho(\Gamma) \rightarrow \rho_e(\Gamma)[1 + \mathbf{A}(\Gamma) \cdot \nabla T(\mathbf{r}) + B_{ij}(\Gamma)\partial_i U_j(\mathbf{r})]. \tag{6}$$

The phase functions $\mathbf{A}$ and B_{ij} are obtained from the formal solution to the Liouville equation:

$$\mathbf{A}(\Gamma) = k_B \beta^2 \left\{ \int d\mathbf{r}'(\mathbf{r}' - \mathbf{r}) - \left[u(\mathbf{r}'; \Gamma) - \frac{h_e}{n_e} n(\mathbf{r}'; \Gamma) \right] - \int_{-\infty}^{0} dt \left[\mathbf{S}(t, \Gamma) - \frac{h_e}{n_e m} \mathbf{P}(t, \Gamma) \right] \right\} \tag{7}$$

$$B_{ij}(\Gamma) = \beta \int d\mathbf{r}'(\mathbf{r}' - \mathbf{r})_i p_j(\mathbf{r}'; \Gamma)$$

$$+ \beta \int_{-\infty}^{0} dt \left[T_{ij}(t, \Gamma) - \delta_{ij} \left\{ p_e V + \frac{\partial p_e}{\partial u_e} \tilde{H}(\Gamma) + \frac{\partial p_e}{\partial n_e} \tilde{N}(\Gamma) \right\} \right]. \tag{8}$$

Here n, u, and p are the phase functions corresponding to the local number, energy, and momentum densities. Their volume integrals are the constants of the motion N, H, and P. The tilde over these quantities represents the deviation from their equilibrium average values, where the latter are denoted with a subscript e. The fluxes of energy and momentum, $\mathbf{S}(t, \Gamma)$ and $T_{ij}(t, \Gamma)$, have arguments indicating that these phase functions are to be evaluated at the phase point $\Gamma(t)$ obtained from Hamilton's equations using the initial condition $\Gamma(0) = \Gamma$, e.g., $\mathbf{S}(t, \Gamma) = \mathbf{S}(\Gamma(t))$. Equation (6) is the familiar linear response ensemble from which the Green-Kubo expressions for thermal conductivity and viscosities follow directly by calculating the average fluxes of energy and momentum. The use of this ensemble to calculate correlations has had much less attention [4].

To simplify the discussion of correlations in linear response we specialize to the case of uniform shear flow. This is a state for which the temperature is uniform but the x component of the velocity field varies linearly in the y direction, $U_x = ay$. The shear rate, a, is the control parameter measuring the deviation from equilibrium and is assumed to be small for the discussion in this section. In this case, substitution of (6) into (1) gives

$$C(\mathbf{r}, \mathbf{r}'; t) \rightarrow C_e(\mathbf{r} - \mathbf{r}') - a\beta \int_0^{\infty} dt \langle \delta n(\mathbf{r}, t)\, \delta n(\mathbf{r}', t) T_{xy} \rangle_e . \tag{9}$$

The first term on the right side of (8) has not been retained since it is similar to contributions discussed for the information entropy ensemble and does not contribute to long-range correlations. Also, the second term on the right side of (9) has been transformed using the time translational invariance of the equilibrium ensemble. This term describes any possible long-range correlations. Clearly it involves the dynamics of fluctuations in the steady state. To estimate the dynamics, note that the invariance of the equilibrium ensemble to space translations implies that the correlation function depends only on $\mathbf{r} - \mathbf{r}'$. A Fourier transform in this variable then gives

$$\tilde{C}(\mathbf{k}; t) - \tilde{C}_e(\mathbf{k}) \rightarrow -a\beta \int_0^\infty dt \, \langle \delta\tilde{n}(\mathbf{k}, t)\delta\tilde{n}(-\mathbf{k}, t)T_{xy}\rangle_e \,. \tag{10}$$

This expression is qualitatively different from (4) since it involves the dynamics of correlations. The long wavelength behavior can be approximated by assuming the dynamics at this space scale is determined from the macroscopic hydrodynamic equations (the Landau-Placzek approximation). Then the density fluctuations $\delta\tilde{n}(\mathbf{k}, t)$ are determined from the linearized hydrodynamic equations

$$\delta\tilde{n}(\mathbf{k}, t) = -\frac{n_e}{c_0} e^{-\Gamma k^2 t} \sin(c_0 k t) i\, \hat{\mathbf{k}} \cdot \delta\mathbf{U}(\mathbf{k}, 0) + \ldots \,, \tag{11}$$

where c_0 is the sound velocity, Γ is the sound damping constant, and the dots indicate contributions from initial perturbations proportional to $\delta\tilde{n}$ and $\delta\tilde{T}$. The latter give vanishing contributions to the correlation function in (10). Use of this result in (10) and inversion of the transform gives the desired asymptotic result

$$\tilde{C}(\mathbf{k}; t) - \tilde{C}_e(\mathbf{k}) \rightarrow -a\beta \left(\frac{n_e}{c_0}\right)^2 \langle (\hat{\mathbf{k}} \cdot \delta\mathbf{U})^2 T_{xy}\rangle_e \int_0^\infty dt \, e^{-2\Gamma k^2 t} \sin^2(c_0 k t). \tag{12}$$

The equilibrium average and the time integral are easily performed, leading to a result that behaves as k^{-2} for long wavelengths. The corresponding real space correlation function behaves asymptotically as

$$C(\mathbf{r}; t) \sim -\left(\frac{n^3 m C_v \chi_T}{4\pi\Gamma C_p \beta}\right) a \frac{\hat{\mathbf{x}}\hat{\mathbf{y}}}{r} \,, \tag{13}$$

where C_p and C_v are the specific heats at constant pressure and volume, and χ_T is the isothermal compressibility. Equation (10) shows that a hydrodynamic spectrum for density fluctuations leads to long-range algebraic decay. In the present case this is due to the correlation of two sound modes in the presence of the externally driven shear field. The hydrodynamic modes are essential for the mode coupling phenomena since the dynamics persists on time scales $\sim\Gamma k^2$ which can be arbitrarily long for long wavelengths, and which generate the k^{-2} singular behavior of $\tilde{C}(\mathbf{k}; t)$ at long wavelengths. Although the analysis of the correlation function in (10) has been phenomenological here, a more detailed many-body calculation leads to the same result [4].

Linear response is valid only to leading order in the shear rate a. More generally, it can be seen that the correlation function has the form

$$l^3[C(\mathbf{r};t) - C_e(\mathbf{r})] = \frac{l}{r}F\left(\frac{r}{l}\right), \quad l = \sqrt{\frac{\Gamma}{a}} \tag{14}$$

where l is the new correlation length introduced by the nonequilibrium state, and $F\left(\frac{r}{l}\right)$ is a scaling function. The linear response result (10) follows from

$$\frac{l}{r}F\left(\frac{r}{l}\right) \to \frac{l}{r}F(0).$$

Interestingly, the perturbation expansion in the shear rate leads to long-range behavior but this should be understood as applying for $r \prec l$; a crossover to some different long-range form is expected for $r \succ l$. Such behavior is discussed in [5] using a Langevin description. Section 3 describes a more complete kinetic theory formulation of this nonperturbative analysis for the special case of a low-density gas for arbitrary nonequilibrium states near or far from equilibrium.

3. KINETIC THEORY OF FLUCTUATIONS

The transport properties of a low-density gas for states near or far from equilibrium can be obtained from the nonlinear Boltzmann equation for phase space density $f(y,t)$. Here $y \leftrightarrow (\mathbf{r}, \mathbf{v})$ is a point in the six-dimensional phase space. A related equation can be derived for the fluctuations, $C(y,y';t)$, under the same conditions of low density, from which the density autocorrelation function can be obtained by integration over the velocities:

$$f(y,t) = \langle \hat{f}(y);t\rangle, \quad C(y,y';t) \equiv \langle \delta\hat{f}(y)\delta\hat{f}(y');t\rangle, \quad \hat{f}(y) = \sum_{i=1}^{N}\delta(y - x_i). \tag{15}$$

$$C(\mathbf{r},\mathbf{r}';t) = \int d\mathbf{v}\, d\mathbf{v}' C(y,y';t). \tag{16}$$

The kinetic equations [6] are somewhat simpler in terms of the pair correlation function, $G(y_1, y_2; t) \equiv C(y_1, t; y_2, t) - \delta(y_1 - y_2)f(y_1, t)$,

$$\left(\frac{\partial}{\partial t} + \mathbf{v}\cdot\nabla\right)f(y,t) = J(y|f(t)), \tag{17}$$

$$\left(\frac{\partial}{\partial t} + \mathcal{L} + \mathcal{L}'\right)G((y,y';t) = T(y_1,y_2)f(y_1,t)f(y_2,t). \tag{18}$$

The nonlinear Boltzmann collision operator, $J(y|f(t))$ and the bi-linear operator $\mathcal{L}$ are given in terms of the binary collision operator $T(y, y_1)$:

$$J(y|f(t)) = \int dy_1 T(y, y_1) f(y, t) f(y_1, t), \tag{19}$$

$$\mathcal{L}h(y) = \mathbf{v} \cdot \nabla h(y) - \int dy_1 T(y, y_1)[h(y, t) f(y_1, t) + f(y, t) h(y_1, t)], \tag{20}$$

where $T(y_1, y_2)$ describes velocity changes due to completed scattering of a pair of particles in contact (at the same space point),

$$T(y_1, y_2) = \delta(\mathbf{r}_1 - \mathbf{r}_2) \int d\hat{\mathbf{r}}\, \Theta\, (\hat{\mathbf{r}} \cdot \mathbf{g}_{12}) \hat{\mathbf{r}} \cdot \mathbf{g}_{12}(b_{12} - 1). \tag{21}$$

Here $\mathbf{g}_{12} \equiv \mathbf{v}_1 - \mathbf{v}_2$ and the scattering operator b_{12} transforms functions of $\mathbf{v}_1$ and $\mathbf{v}_2$ to their precollision velocities, i.e., $b_{12}X(\mathbf{v}_1,\mathbf{v}_2) = X(\mathbf{v}_1{}',\mathbf{v}_2{}')$ with $\mathbf{v}_1{}' = \mathbf{v}_1 - (\hat{r} \cdot \mathbf{g}_{12})\hat{r}$ and $\mathbf{v}_2{}' = \mathbf{v}_2 - (\hat{r} \cdot \mathbf{g}_{12})\hat{r}$. There is a close relationship of the equation for the correlations to the nonlinear Boltzmann equation. The dynamics of the former is governed by $\mathcal{L}$ which is the linearization of the Boltzmann operator about the instantaneous nonequilibrium state. Furthermore, the source of correlations on the right side of (18) is precisely the binary collision operator defining the nonlinear Boltzmann operator.

To discuss long-range correlations it is useful to consider the special case of stationary states (perhaps including external forces and boundary conditions). The formal solution to (18) is

$$G(y, y'; t) = (\mathcal{L} + \mathcal{L}')^{-1} T(y, y') f(y, t) f(y', t). \tag{22}$$

Substitution of this result into (16) gives

$$C(\mathbf{r}, \mathbf{r}'; t) = n(\mathbf{r})\delta(\mathbf{r}_1 - \mathbf{r}_2) + \int d\mathbf{v} d\mathbf{v}' (\mathcal{L} + \mathcal{L}')^{-1} T(y, y') f(y, t) f(y', t). \tag{23}$$

The second term can generate long-range correlations if $\mathcal{L}$ contains hydrodynamic modes in its spectrum. This is certainly the case for simple fluids, as follows from the local conservation laws. However, at strict equilibrium detailed balance implies

$$T(y, y') f_e(y) f_e(y') = 0. \tag{24}$$

Thus, for a system with local conservation laws there are hydrodynamic modes with excitations that decay on arbitrarily long-time scales for long wavelengths. Equation (23) implies there will be long-range correlations *generically* except in the special case of equilibrium. Surprisingly, long-range correlations are the norm while short-range equilibrium correlations are the exception.

4. DISCUSSION

The above analysis has identified a mechanism for long-range correlations in nonequilibrium states. It is a dynamical mechanism related to the existence of collective modes whose relaxation times grow with the wavelength of the spatial perturbation (hydrodynamic modes). The simple perturbation result (9) shows this clearly where the long-range effects are due to dynamic correlations of the spontaneous density fluctuations in the system with perturbations of the equilibrium state, and follow directly from the existence of hydrodynamic excitations. These long-range effects do not require any significant interatomic force range nor any external long-range forces. The kinetic theory result (18) illustrates this most clearly since the source for correlations, $T(y_1, y_2)f(y_1, t)f(y_2, t)$, is proportional to $\delta(\mathbf{r}_1 - \mathbf{r}_2)$ yet any finite amplitude leads to long-range correlations through the hydrodynamic spectrum of $\mathcal{L}$ in (23). This amplitude exists for all states $f(y_1, t)$ *except* equilibrium where it vanishes due to detailed balance. Thus it appears that sufficient conditions for long-range correlations are the existence of local conserved densities leading to hydrodynamic modes, and the violation of detailed balance. The latter is most easily generated by boundary conditions driving the system to inhomogeneous stationary states. A rather complete description of boundary conditions for this problem in the context of a Langevin description is given in [7]. However, it is also possible to have a violation of detailed balance for spatially homogeneous states [8].

A convincing verification of this mechanism for long-range correlations has been given for a class of *lattice gas cellular automata* (LGA). These are systems of N particles whose dynamics is defined on a lattice in space and time. Local collision rules are defined for each node of the spatial lattice constrained by the conservation laws for mass and momentum and such that the dynamics on large space and time scales reduces to the fluid dynamics equations. These conditions assure the existence of hydrodynamic modes. Violation of detailed balance can be accomplished in this context by having stochastic collision rules with the probabilities for some direct and inverse collisions to be different. This is an example of a homogeneous condition for a nonequilibrium state. It is possible to derive a kinetic theory for the LGA analogous to that of Section 3 [9]. Detailed numerical solution to the coupled equations for the distribution function and correlations has verified in detail the generation of long-range correlations from these local collision rules.

Most theoretical analyses, and the experiments of Sengers *et al.*, apply for states near equilibrium. The kinetic theory description of Section 3 can be applied more generally to hydrodynamic excitations about states far from equilibrium. Recently, this was done for small spatial perturbations around uniform shear flow allowing for arbitrary values of the shear rate in the reference state [10]. The long-range hydrodynamic modes and associated correlations become functions of the shear rate far from equilibrium. An unexpected result of this study was the discovery of a long wavelength instability [11]. Thus, correlations are generated up to a critical wavelength above which the reference state is unstable. At these larger wavelengths the instability leads to formation of a completely different, inhomogeneous and time dependent state. This illustrates the additional complexity of long-range correlations to be expected for states far from equilibrium.

Finally, we mention recent studies of models for granular flow. The simplest is a system of N smooth hard spheres with inelastic collisions. The fact that energy is lost on each collision implies a violation of detailed balance and the associated Boltzmann equation for an isolated system does not support an equilibrium state. However, there is a uniform state describing a uniform cooling due to the inelasticity, and it is possible to show that hydrodynamic modes exists for spatial fluctuations about this uniform cooling state. The source term for correlations in (18) does not vanish for this state and hence long-range correlations will result, again complicated by the existence of instabilities. The details of these correlations are under investigation.

REFERENCES

1. LI, W.B. , P.N. SEGRE, R.W. GAMMON & J.V. SENGERS. 1994. Small-angle Raleigh Scattering from Nonequilibrium Fluctuations in Liquid Mixtures. Physica A **204:** 399–436.
2. DORFMAN, J.R., T.R. KIRKPATRICK & J.V. SENGERS. 1994. Generic long-range correlations in molecular fluids. Ann. Rev. Phys. Chem. **45:** 213–239.
3. MCLENNAN, J.A. 1989. Introduction to Nonequilibrium Statistical Mechanics. Prentice Hall. Engelwood Cliffs, NJ.
4. DUFTY, J.W. 1981. Light scattering from a nonequilibrium fluid. *In* Spectral Line Shapes, B. Wende, Ed.: 1143–1158. W. de Gruyter. New York.
5. LUTSKO, J. & J.W. DUFTY. 1985. Hydrodynamic fluctuations at large shear rate. Phys. Rev. A. **32:** 3040–3054.
6. DUFTY, J.W., M. LEE & J.J. BREY. 1995. Kinetic model for pair correlations. Phys. Rev. E. **51:** 297–309; MARCHETTI, M.C. & J.W. DUFTY. 1983. Kinetic and hydrodynamic theories of nonequilibrium fluctuations. Physica **118A:** 205–218; Ernst, M.H. & E.GD. Cohen. 1981. Nonequilibrium fluctuations and the hierarchy. *In* Perspectives in Statistical Physics. Raveche, H.J., Ed. North Holland. Amsterdam.
7. SCHMITZ, R. 1988. Fluctuations in nonequilibrium fluids. Phys. Rep. **171:** 1–58.
8. PAGONABARRAGA, I. & J.M. RUBI. 1994. Long-range correlations in diffusive systems away from equilibrium. Phys. Rev. E. **49:** 267–272.
9. BUSSEMAKER, H.J., M.H. ERNST & J.W. DUFTY. 1995. Generalized Boltzmann equation for lattice gas automata. J. Stat. Phys. **78:** 1521–1554.
10. LEE, M. & J.W. DUFTY. 1997. Transport far from equilibrium — uniform shear flow. Phys. Rev. E. **56:** 1733.
11. LEE, M., J.W. DUFTY, J.M. MONTANERO, A. SANTOS & J.F. LUTSKO. 1996. Long wavelength instability for uniform shear flow. Phys. Rev. Lett. **76:** 2702.

Renormalization of Self-Organized Critical Models

JEFF HASTY AND KURT WIESENFELD

School of Physics, Georgia Institute of Technology, Atlanta, Georgia 30332

ABSTRACT: Self-organized criticality refers to a mechanism whereby complex dissipative dynamical systems naturally evolve to a self-sustaining critical point, exhibiting correlations on all length scales and time scales. A recent theory based on a real space renormalization group approach sheds light on fundamental aspects of the phenomenon.

1. SELF-ORGANIZED CRITICALITY

Many phenomena can be explained theoretically by isolating a specific scale of interaction. This characteristic length or time scale is of great utility in that one can ignore or treat perturbative effects which are thought to occur on other scales. For example, the nuclear interaction can be ignored if one seeks to explain some atomic phenomenon such as chemical bonding. Thus one can speak of correlations which become vanishingly small and are therefore negligible. In dynamical systems, *scale invariance* refers to the antithesis of this picture. Scale invariant phenomena have no characteristic scale of interaction; small perturbations can propagate throughout the entire system and cannot be ignored.

The physical motivation for self-organized criticality (SOC) lies in the belief that there is a large class of phenomena where scale invariance arises naturally; that is, the underlying dynamical process leads to scale invariant observables. Couched in the language of equilibrium systems, the critical state is obtained without the fine-tuning of a parameter.

The prototypical model for SOC is known as the BTW sandpile model, which was introduced nearly a decade ago in the first work on the subject [1]. On an $L \times L$ lattice, the cellular automaton (CA) model is defined by two simple updating rules. First the pile is driven by randomly choosing a site (i, j) and dropping a grain of "sand" by updating the height h_{ij},

$$h_{i,j} \rightarrow h_{i,j} + 1. \qquad (1.1)$$

Secondly, if the local height exceeds the threshold value h_c, the site relaxes by distributing sand to its nearest neighbors,

$$\begin{aligned} h_{i,j} &\rightarrow h_{i,j} - 4 \\ h_{i\pm 1,j} &\rightarrow h_{i\pm 1,j} + 1 \\ h_{i,j\pm 1} &\rightarrow h_{i,j\pm 1} + 1. \end{aligned} \qquad (1.2)$$

This relaxation can then cause neighboring sites to become excited (above threshold), thus triggering a sequence of relaxation events known as an avalanche. A separation of time scales is built into the simulation; dropping does not take place until all sites are relaxed. Sand leaves the system by "falling off" at the boundary.

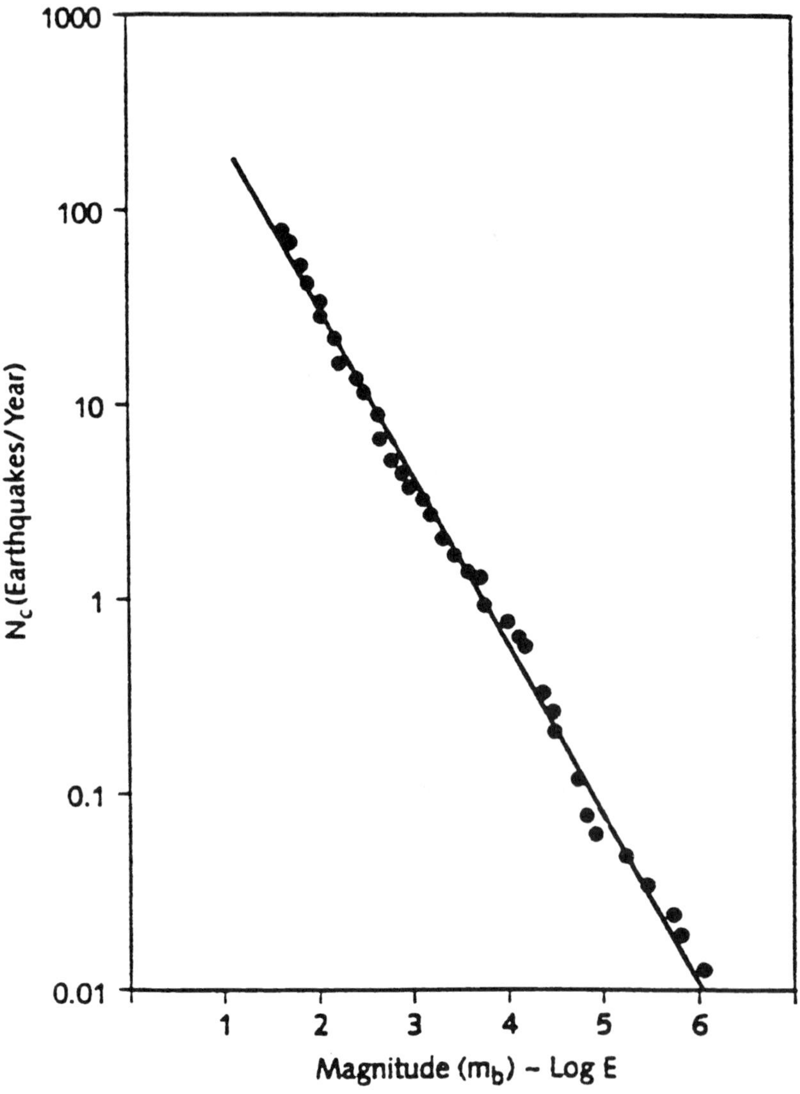

FIGURE 1. Earthquake size distribution along the New Madrid fault zone from 1973–1983 (reproduced from [5]).

This model, which is intended more as a paradigm than a real simulation of sand, provides a framework for studying certain driven-diffusive systems. Independent of initial conditions, the simulations evolve through a transitory period to a state characterized by power laws in the statistical distributions of certain observables. For example, the avalanche size is defined as the number of sites which relax in a single avalanche, with distribution given by

$$D(s) \sim s^{1-\tau}, \qquad (1.3)$$

where $D(s)$ is the probability that s or more sites are involved.

There are of course numerous spatial and temporal measurements which can be made, and there has been considerable work in the direction of deducing various scaling laws which relate their distribution exponents [2]. Here, however, we will maintain a narrow focus, concentrating solely on the avalache size distribution.

Self-organized criticality has been invoked to explain many diverse phenomena, with earthquakes being perhaps the most widely cited example [3]. An empirical result describing the distribution of earthquake sizes is the well established Gutenberg-Richter law [4]. In FIG. 1, the earthquake size distribution obtained from 1973–1983 along the New Madrid fault zone in the central United States is plotted [5]. The distribution obeys $E(m) \sim m^{-\beta}$, where $E(m)$ is the number of earthquakes of size m or greater. That such a simple fit would arise is remarkable, but perhaps even more astounding is the claim that this distribution is obeyed in any geographical region, with the only region-specific quantity being the value of the exponent β.

A simple model used by geophysicists to describe one class of earthquakes is that of two plates shearing against one another. In this picture, plate tectonics slowly drives one plate against another until some local position exceeds a static friction threshold and slips. This slipping can then cause neighboring areas to slip, and an avalanche of slipping events is an earthquake. In 1967 geophysicists Burridge and Knopoff [6] proposed a spring-block model for this type of plate dynamics. In their model, a fixed plate supports a two-dimensional array of blocks connected by springs to each other and to a moving plate (see FIG. 2.).

In 1992 Chirstensen and Olami [7] analyzed the Burridge-Knopoff model with an eye towards SOC. They were able to show how a dissipative sandpile model can be derived from the equations of motion by using a threshold description of the friction-

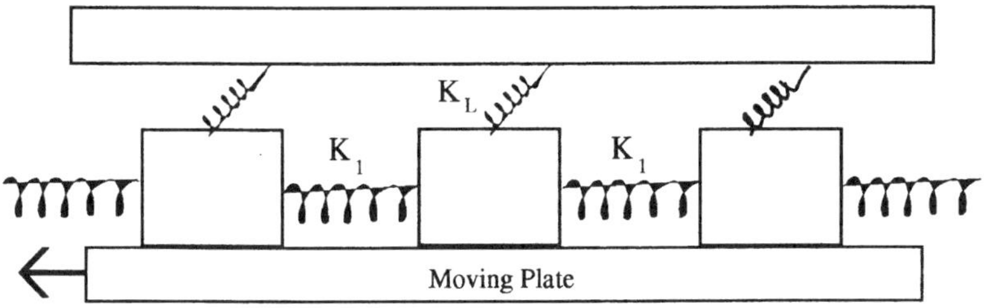

FIGURE 2. Schematic of the Burridge-Knopoff spring-block model.

al force and a separation of time scales. For example, if one lets the displacement of each block in FIG. 2 be denoted $x_{i,j}$, then the total force on block (i, j) is given by

$$F_{i,j} = K_1(2x_{i,j} - x_{i-1,j} - x_{i+1,j}) + K_2(2x_{i,j} - x_{i,j-1} - x_{i,j+1}) + K_L x_{i,j}, \qquad (1.4)$$

where the K's denote the elastic constants for the springs. Now the general idea is to let the plates slide slowly against each other until one site reaches some threshold and slips, triggering the process of relaxation. Christensen and Olami show that the redistribution of force after a local slip at position $x_{i,j}$ is given by

$$\begin{aligned} F_{i\pm1,j} &\to F_{i\pm1,j} + \delta F_{i\pm1,j} \\ F_{i,j\pm1} &\to F_{i,j\pm1} + \delta F_{i,j\pm1} \\ F_{i,j} &\to 0, \end{aligned} \qquad (1.5)$$

where the increase in the force on the nearest-neighbor blocks is

$$\begin{aligned} \delta F_{i\pm1,j} &= \frac{K_1}{2K_1 + 2K_2 + K_L} F_{i,j} = \alpha_1 F_{i,j} \\ \delta F_{i,j\pm1} &= \frac{K_2}{2K_1 + 2K_2 + K_L} F_{i,j} = \alpha_2 F_{i,j}. \end{aligned} \qquad (1.6)$$

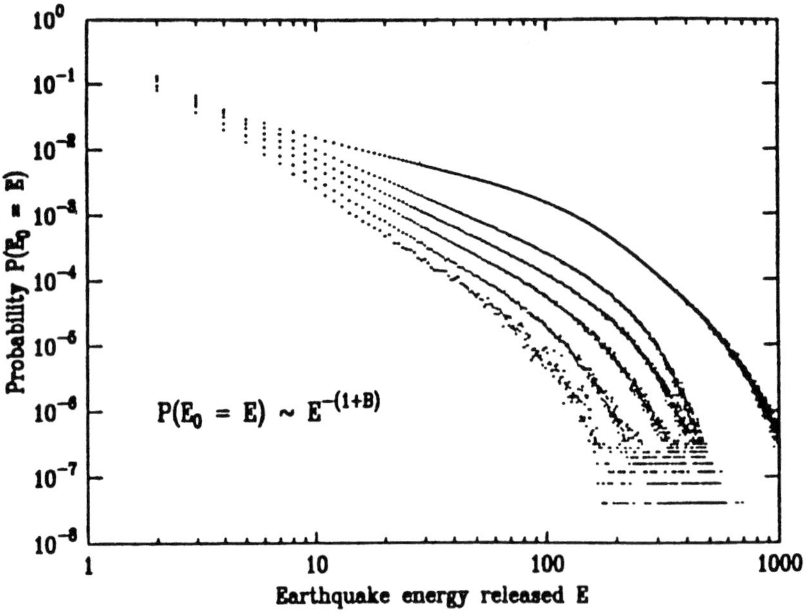

FIGURE 3. Simulation results of the Christensen-Olami model (reproduced from [7]). Plotted is the distribution for the number of slips in an avalanche (energy released in an earthquake), with the differing curves representing differing amounts of dissipation.

Next an algorithm for simulating the system is generated by exploiting a supposed separation of time scales; one assumes that in real systems the relaxation process (earthquakes) are orders of magnitude faster than the driving (plate motion), so that the system need only be driven when all sites are stable. The simulation can then be summarized as follows: A random site is chosen and a small amount of force is added. When the force at a given site exceeds some threshold value F_{th}, this site will relax according to (1.5). This in turn can cause the neighboring sites to exceed F_{th} so that they then may relax to their neighbors, and so on. This relaxation event or earthquake continues until no sites exceed F_{th}, whence the driving continues by choosing a site randomly as before.

This model is very similar to the BTW model previously described, with the notable difference being the possible lack of local conservation; for $\alpha_1 = \alpha_2 = .25$ the model is locally conservative, but for other values of α the model is not. The simulation results reproduce the Guttenburg-Richter law, yielding differing exponents depending on the degree of local conservation (see FIG. 3). This result is quite interesting in light of empirical data which suggests that the exponential value is a function of the media, and one could argue heuristically that the K values (and hence the α values) should depend on the geographic region.

2. A RENORMALIZATION THEORY FOR SOC

Although CA models such as those explained above are easy to simulate, a general framework for deducing statistical distributions from them has not been fully developed. In this section we explain a recently proposed renormalization scheme capable of calculating distribution exponents and explaining why different models can exhibit the same exponent values (universality).

The Pietronero, Vespignani, Zapperi (PVZ) [8] approach uses a real space coarse-graining of the local dynamics to yield a renormalization group (RG) procedure. The original implementation was in the context of the BTW model described above and a similiar model known as the Manna model [9], [10]. The Manna model differs from the BTW model only in its relaxation rule; an excited site chooses two neighbors at random for distribution. Both models evolve to a stationary state characterized by power law distributions for various observables. In the SOC state, the Manna simulation yields an avalanche exponent of $\tau = 1.28$ [10], close to the value of 1.22 obtained for the BTW model [1].

The first step in the PVZ theory is to define a space of relaxation probabilities, whereby each point in the space corresponds to a specific model. This space is spanned by the four probabilities P_j ($j = 1, 2, 3, 4$), with a point in the space described by the vector $\mathbf{P} = (P_1, P_2, P_3, P_4)$. Within this framework, the BTW and Manna models correspond to $\boldsymbol{P} = (0, 0, 0, 1)$ and $P = (0, 1, 0, 0)$, since the probability that the BTW model relaxes by distributing sand to four neighbors is unity, while the Manna model relaxes to two neighbors with probability one.

The next step is to calculate how a given point evolves under a change of scale. PVZ first define a 2×2 block of sites with one site excited initially (above threshhold), and then find all relaxation paths generated from the excited block. How each path affects its neighboring blocks is monitored, and a sum is taken over the paths leading to a given block relaxation. This is done for all possible initial block configurations (a block configuration being composed of internal threshhold and below-threshhold sites), and a weighted average is taken. The weight factor is determined by the probability of obtaining a given initial block, and this involves an additional parameter ρ which defines the density of threshhold sites. ρ is itself determined from **P** via the *balance condition*,

$$1 = \rho(P_1 + 2P_2 + 3P_3 + 4P_4), \tag{2.1}$$

which effectively defines ρ so that the local conservation rule (sand in = sand out) is satisfied on every scale.

This procedure yields a map of the form **P′**(**P**), where **P′** denotes the block relaxation probabilities. The map is then iterated with either the BTW or the Manna initial conditions, and a unique fixed point common to both models is found. This fixed point determines the exponent τ (cf. Eq. (1.3)), and PVZ obtain a value of τ = 1.253, in good agreement with the simulation results cited above.

3. RENORMALIZATION OF A DIRECTED SANDPILE MODEL

It is impressive that this approach generates the right exponent τ from first principles. However, the RG picture predicts much more, and its success or failure should rest on a systematic check of all of its predictions. In this section we explain the application of a PVZ-type RG scheme to a model known as the directed sandpile model [11]. This model, while not as complex as the BTW or Manna models, enables us to explore the implications of the RG theory, and to test its validity. Further, there are known analytic results [11] for the avalanche exponents.

The directed model is defined as follows. The system is perturbed by randomly choosing one of the sites on the upper row of the lattice and updating;

$$h_{1,j} \to h_{1,j} + 1. \tag{3.1}$$

Next, if the height h at any site exceeds 1, then the site relaxes by transferring one grain to each of its two forward neighbors:

$$\begin{aligned} h_{i,j} &\to h_{i,j} - 2 \\ h_{i+1,j\pm} &\to h_{i+1,j\pm} + 1, \end{aligned} \tag{3.2}$$

where $j\pm = j \pm [1 \pm (-1)^i]/2$. We will refer to this model as the DR model.

Dhar and Ramaswamy [11] have shown analytically that, in the scaling region, the avalanche distribution is given by $P(s) \sim s^{-4/3}$, where $P(s)$ is the probability that s sites tumble in a single event.

As before, the RG scheme specifies how the rule (3.2) transforms under a change of scale. Mathematically, it provides a mapping on the space where each point represents a particular CA rule.

The explicit application involves the summing of all internal relaxation paths which lead to a given type of diamond relaxation, and the details will be provided elsewhere. The end result is that an RG operator is derived which transforms the dynamics of inter-diamond sites (which are themselves diamonds) to a dynamics for the coarse-grained diamond cell. The space on which the RG operator acts is thus a

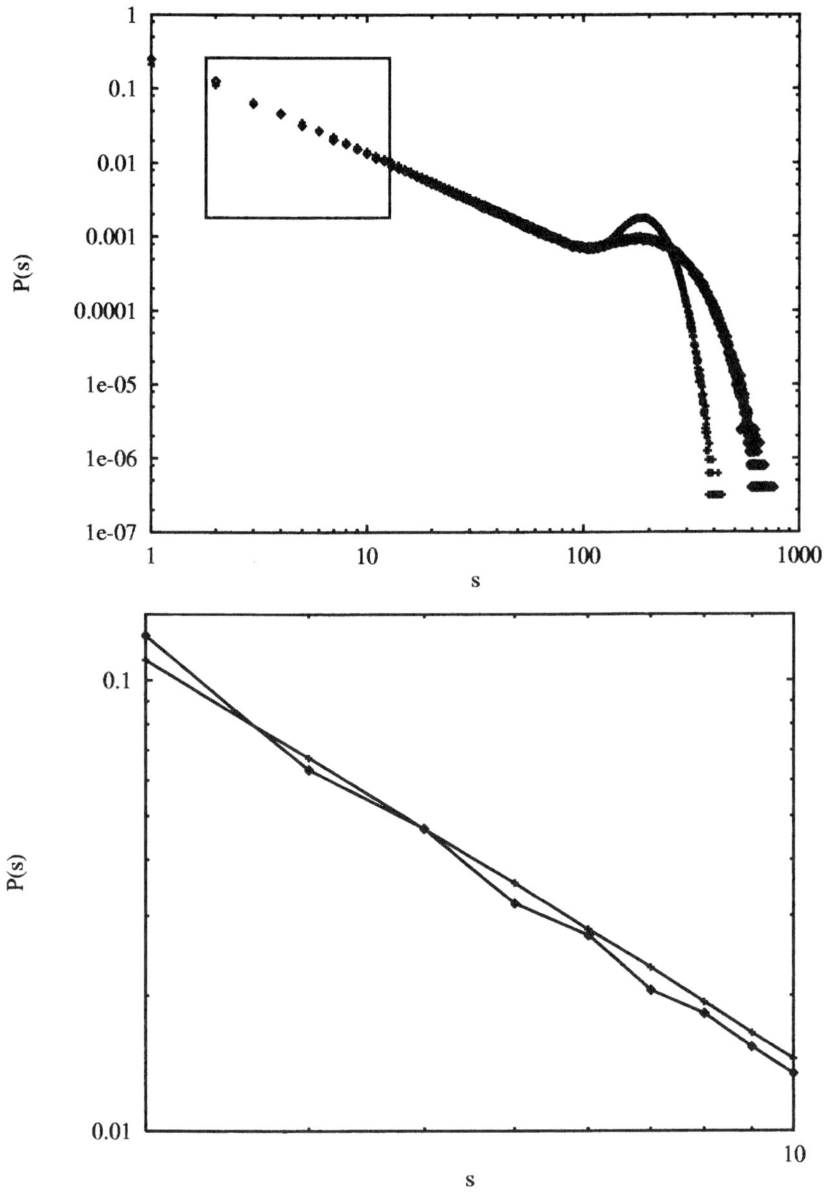

FIGURE 4. Avalanche distribution for the model corresponding to the fixed point of the RG map (+) and the original DR model data (o). The small length scale blow up highlights the power law behavior of the fixed point model.

space where each point represents a particular cellular automata rule, with the directed model being one example.

The RG map can then be analyzed, and the results checked against the analytic results or by using simulations. For the initial condition rule corresponding to the DR model, the RG map is attracted to a fixed point. This fixed point implies that there is indeed scale invariance, and we can calculate the avalanche exponent from it. We obtain $\tau = 1.325$, in excellent agreement with the known result. We next simulate the rule corresponding to this fixed point and measure the avalanche distribution (FIG. 4). This data gives an exponent of $\tau = 1.32$, in the same class as the directed model as predicted by the theory. Further, the small length scale power law deviations in the directed model data have disappeared, also as predicted by the RG picture: the DR model does not correspond to a fixed point and therefore we do not expect scale invariance at small scales, while the fixed point model should be invariant at all scales.

The RG theory predicts the existence of universality classes corresponding to the basin of attraction of each fixed point; all models in the basin should share a common avalanche exponent. We find that the above-mentioned fixed point has a large basin of "symmetric rules" described as follows. If we let R/L be the probability that a site relaxes by distributing sand to its lower right/left neighbors, then symmetric rules are those which have $R = L$. Preliminary simulation results for the avalanche distributions for various symmetric rules yield a common exponent for all models as predicted. Additionally, we find that the further a rule is from the fixed point (as measured by the number of iterates), the larger the region of small avalanche power law deviation, again consistent with the RG picture.

4. SUMMARY

The RG implementation presented above provides valuable information about SOC models. Further generalizations could yield still more. For example, the size of the system could be included in the analysis, potentially explaining finite size effects such as the exodus from the scaling region in data such as FIG. 4. Another unexplored area is the transient dynamics as the system evolves toward the SOC state, and it may be possible to learn something of the self-organizing behavior by exploiting the RG method.

There are also limitations to the current RG approach as it is currently applied. In particular, coarse-graining inevitably throws out correlations inherent on the attractor [12], and there is as yet no systematic method for including higher order corrections to the RG map. Potentially, these are only technical points provided the overall structure of the theory hangs together, as our work on the directed models indicates. Another open question is whether a complete RG theory will be able to explain the nonuniversal behavior seen in the Christensen-Olami model.

We have described one way in which cellular automata are used to model nonequilibrium phenomena, and how SOC can arise as a consequence. In modeling such "complex systems," the specification of all relevant details yields models which are often beyond analysis. Universality, in this respect, is a trump card. If one can demonstrate that a large number of models have common features, then one may proceed by analyzing the simplest of the models. The power of a renormalization theory lies not only in its ability to deduce critical exponents, but also in its explanation and prediction of such universality classes.

REFERENCES

1. P. BAK, C. TANG & K. WIESENFELD. 1987. Self-organized criticality: An explanation of 1/f noise. Phys. Rev. Lett. **59:** 381–384; 1988. Self-organized criticality. Phys. Rev. A **38:** 364–374.
2. C. TANG & P. BAK. 1988. Mean field theory of self-organized critical phenomena. J. Stat. Phys. **51:** 797–802; K. CHRISTENSEN, H.C. FOGEDBY & H.J. JENSEN. 1991. Dynamical and spatial aspects of sandpile cellular automata. J. Stat. Phys. **63:** 653–684.
3. P. BAK & C. TANG. 1989. Earthquakes as a self-organized critical phenomenon. J. Geophys. Res. B. **94:** 15635–15637; A. SORNETTE & D. SORNETTE. 1989. Self-organized criticality and earthquakes. Europhys. Lett. **9:** 197–202; K. CHEN, P. BAK & S.P. OBUKHOV. 1990. Self-organized criticality in a crack-propagation model of earthquakes. Phys. Rev. A. **43:** 625–630; K. CHRISTENSEN & Z. OLAMI. 1992. Variation of the Gutenberg-Richter *b* values and nontrivial temporal correlations in a spring-block model for earthquakes. J. Geophys. Res **97:** 8729–8735.
4. B. GUTENBERG & C.F. RICHTER. 1956. Magnitude and energy of earthquakes. Ann. di Geofis. **57:** 1–15.
5. A.C. JOHNSTON & S. NAVA. 1985. Recurrence rates and probability estimates for the New Madrid seismic zone. J. of Geophys. Res. **90:** 6737–6753.
6. R. BURRIDGE & L. KNOPOFF. 1967. Model and theoretical seismicity. Bull. Seismol. Soc. Am. **57:** 341–371.
7. K. CHRISTENSEN & Z. OLAMI. 1992. Variation of the Gutenberg-Richter *b* values and nontrivial temporal correlations in a spring-block model for earthquakes. J. Geophys. Res **97:** 8729–8735.
8. L. PIETRONERO, A. VESPIGNANI & S. ZAPPERI. 1994. Renormalization scheme for self-organized criticality in sandpile models. Phys. Rev. Lett. **72:** 1690–1693; A. VESPIGNANI, S. ZAPPERI & L. PIETRONERO. 1995. Renormalization approach to the self-organized critical behavior of sandpile models. Phys. Rev. E. **51:** 1711–1724.
9. S.S. MANNA. 1991. Two-state model of self-organized critical phenomena. J. Phys. A **24:** L363–L369.
10. S.S. MANNA. 1990. Large-scale simulation of avalanche cluster distribution in sand pile model. J. Stat. Phys. **59:** 509–521; 1991. Critical exponents of the sandpile models in two dimensions. Physica A. **179:** 249–268.
11. D. DHAR & R. RAMASWAMY. 1989. Exactly solved model of self-organized critical phenomena. Phys. Rev. Lett. **63:** 1659–1662.
12. J. HASTY & K. WIESENFELD. 1997. Renormalization of one-dimensional avalanche models. J. Stat. Phys. **86:** 1179–1201.

Two-Dimensional Vortex Crystals[a]

D. Z. JIN AND D. H. E. DUBIN

Physics Department, University of California at San Diego, La Jolla, California 92093

ABSTRACT: Magnetically confined pure electron columns evolve as nearly inviscid, incompressible two-dimensional (2-D) fluids. Experiments on these columns have found that the decay of 2-D turbulence leads to spontaneous formation of "vortex crystals," which are symmetric arrays of strong vortices within a background of lower vorticity. These vortex crystals can be described as maximum fluid entropy states for a system with an ergodically mixed background and a conserved number of strong vortices.

1. INTRODUCTION

The study of turbulence in inviscid, incompressible 2-D fluids is important for understanding geophysical and astrophysical flows [1], [2]. The Great Red Spot of Jupiter, for example, has been modeled in terms of inviscid 2-D Euler flow [3]. Magnetically confined pure electron columns are excellent systems for quantitative study of the free relaxation of 2-D turbulence. In a recent experiment on such columns, the vorticity $\omega(\mathbf{r}, t) = \nabla \times \mathbf{v}(\mathbf{r}, t)$ of the turbulent flow relaxed from some initial conditions to a "vortex crystal" equilibrium, which consists of small intense patches of vorticity (strong vortices) forming a stable equilibrium pattern in a low vorticity background [5]. These vortex crystal states have not been previously observed or predicted to form from turbulent initial conditions. Recently, we have shown that the vortex crystals can be described as the maximum fluid entropy states of a system with an ergodically mixed background and a conserved number of strong vortices [6]. In this paper we discuss the experiment and our theory for the vortex crystals.

2. THE EXPERIMENT

The experimental device is shown in FIG. 1. The electron columns are confined radially by a uniform magnetic field, B_z, and contained axially by voltage applied to end sections of the cylindrical walls. The z-averaged electron density $n(r, \theta, t)$ is accurately measured by dumping the column onto a phosphor screen imaged by a 512 $\times$ 512 CCD camera. Time evolution is inferred from a sequence of shots with differing hold times. This is possible because the shot-to-shot variations in the initial profiles can be controlled within small error.

[a]This work was supported by NSF grant PHY94-21318.

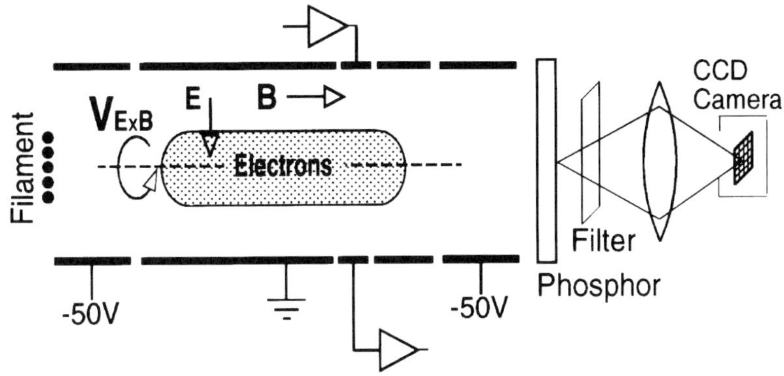

FIGURE 1. The cylindrical experimental apparatus with phosphor screen/CCD camera diagnostic.

In the experiment, the axial bounce frequency of an electron is large compared to the E × B drift rotation frequency; therefore, the evolution of $n(r, \theta, t)$ is described by the 2-D drift-Poisson equation, i.e.,

$$\frac{\partial n}{\partial t} + \mathbf{v} \cdot \nabla n = 0, \quad \mathbf{v} = -\frac{c}{B_z} \nabla \Phi \times \hat{\mathbf{z}}, \quad \nabla^2 \Phi = 4\pi e n. \tag{1}$$

These equations are isomorphic to the Euler equations for an incompressible inviscid 2-D fluid. The measured electron density is proportional to the vorticity of the flow, i.e., $\omega = 4\pi e c n / B_z$, and the electrostatic potential is proportional to the stream function ψ, i.e., $\psi = -c\Phi/B_z$. The nonzero $\partial\Phi/\partial r$ at the wall gives a true free-slip boundary condition.

Vortex crystals form from a highly filamented electron density distribution, as shown in FIG. 2(a). Within one turnover time of the electron column ($\tau_R = 170\ \mu s$), $N_v = 50–100$ strong vortices of roughly equal circulation form due to Kelvin-Helmholtz instabilities, as shown in FIG. 2(b). Subsequently, the turbulent state evolves and relaxes by chaotic vortex advection and mergers, as shown in FIG. 2(c). As a consequence, the number of the strong vortices decreases as $N_v \sim t^{-\xi}$, with $\xi \approx 1^5$. This relaxation is generally consistent with a dynamical scaling theory based on conserved quantities in repeated vortex merger [7].

However, the chaotic vortex motions eventually "cool" down and the vortices no longer merge. The strong vortices then relax into a lattice pattern, which rotates with the background, as shown in FIGS. 2(d) and 2(e). Seemingly, the increase of the number of strong vortices at $600\tau_R$ (FIG. 2(e)) compared to that at $60\tau_R$ (FIG. 2(d)) is because the initial conditions that lead to each image are slightly different owing to the shot noise. The vortex crystal states survive for about $10^4\tau_R$. After this period, individual vortices decay away and the remaining strong vortices adjust to a new rigidly rotating symmetric pattern.

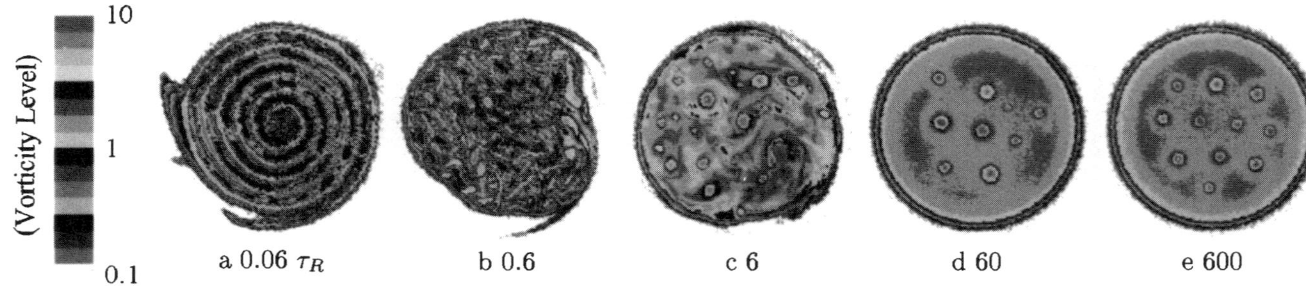

FIGURE 2. Formation of vortex crystal. False color contour plots of vorticity (represented in gray scale here) at five times are displayed. The unit of the vorticity is Γ/r_{ω}^{2}, where r_{ω} is the radius of the conducting boundary.

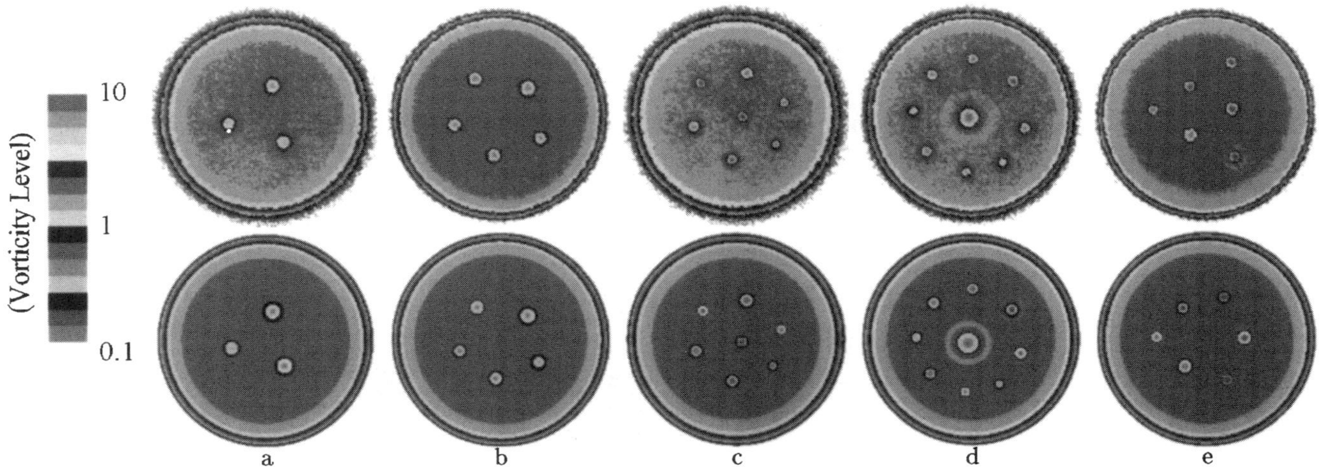

FIGURE 3. Top: examples of experimental images of vortex crystal states. Bottom: corresponding maximum fluid entropy states.

Experimentally, the evolution conserves total circulation Γ, angular momentum L, and energy H. Here

$$\Gamma = \int d\mathbf{r}^2 \omega, \tag{2}$$

$$L = -\int d\mathbf{r}^2 \omega r^2/2, \tag{3}$$

$$H = \int d\mathbf{r}^2 |\mathbf{v}|^2/2 = \int d\mathbf{r}^2 \psi\omega/2. \tag{4}$$

3. MAXIMUM FLUID ENTROPY THEORY

After the chaotic vortex mergers stop, the flow continues to evolve. The random motions of the strong vortices slowly "cool" as the vortices relax toward an equilibrium, and during this evolution the background also evolves. We have found that the final equilibrium of flow, the vortex crystal, can be described as a maximum fluid entropy state of a system with a low vorticity background and a conserved number of strong vortices. Two properties of the final states of the system can be predicted by the maximum fluid entropy theory: the distribution of the background vorticity $\omega_b(\mathbf{r})$, and the positions of the strong vortices $\mathbf{R}_i$, $i = 1, M$. Comparison between the theoretical predictions and the experimental data shows good agreement.

Maximization of the entropy of the system is constrained by the conservation of the total circulation, angular momentum and energy. These constraints follow from the dissipationless nature of ideal fluid flow, together with the cylindrical symmetry and free-slip boundary conditions of the experimental geometry. Also, the incompressibility of vorticity implies that infinitesimal vorticity elements can not overlap during their mixing, which provides another constraint.

The entropy associated with a macroscopic coarse-grained background vorticity distribution $\omega_b(\mathbf{r})$ can be calculated by counting the number of ways of arranging microscopic vorticity elements to obtain the given coarse-grained vorticity. For the simplest case of the microscopic vorticity elements having the same positive vorticity ω_f, the entropy is [8]:

$$S[\omega_b(\mathbf{r})] = -\int d\mathbf{r}^2 \left[\frac{\omega_b(\mathbf{r})}{\omega_f}\ln\frac{\omega_b(\mathbf{r})}{\omega_f} + \left(1 - \frac{\omega_b(\mathbf{r})}{\omega_f}\right)\ln\left(1 - \frac{\omega_b(\mathbf{r})}{\omega_f}\right)\right]. \tag{5}$$

The second term is due to the incompressibility of the vorticity elements, and does not appear in the usual (Boltzmann) expression of the entropy for a compressible flow. The contribution to the entropy from the positions of the strong vortices is negligible because the background has an infinite number of degrees of freedom. Therefore S is also the entropy of the entire (strong vortices + background) system.

The vorticity distribution of the strong vortices are assumed to be described by fixed axially symmetric functions $\omega_{v,i}(|\mathbf{r} - \mathbf{R}_i|)$. Since the vortices are quite intense and small, it does not make a great deal of difference what functional form one chooses for $\omega_{v,i}$ but we choose the (approximately Gaussian) form associated with observed vortices. The stream function ψ is related to the vorticity ω of the flow via Poisson's equation

$$\nabla^2\psi = -\omega. \tag{6}$$

The maximization of S while keeping H, L, and Γ constant is done by the standard technique of Lagrange multipliers: $\delta F = 0$, where $F \equiv S - \beta(H - \Omega L + \mu\Gamma)$, and β, Ω, and μ are Lagrange multipliers that can be interpreted as inverse temperature, rotation frequency, and the chemical potential, respectively. Variation of F with respect to vortex positions $\{\mathbf{R}_i\}$ yields

$$\frac{\partial}{\partial \mathbf{R}_j}(H - \Omega L) = 0. \tag{7}$$

Since $H - \Omega L$ is the Hamiltonian of the system in the rotating frame of frequency Ω, the above equation shows that in the maximum entropy state, the velocity of the strong vortices is zero in this rotating frame: in other words, the strong vortices are in equilibrium, rotating rigidly at frequency Ω. Furthermore, since S is maximized, this equilibrium must be stable.

Variation of F with respect to the background vorticity $\omega_b(\mathbf{r})$ yields the coarse-grained background vorticity:

$$\omega_b(\mathbf{r}) = \frac{\omega_f}{e^{\beta\omega_f\phi} + 1}, \tag{8}$$

where $\phi \equiv \psi + \frac{1}{2}\Omega r^2 + \mu$ is the stream function in the rotating frame. This is very similar to the Fermi distribution in quantum statistics. Equations (7) and (8) characterize the maximum fluid entropy states of our system.

In order to test whether experimentally observed vortex crystal states can be explained by the maximum fluid entropy states, the predictions of the theory are compared with the experimental data. The maximum fluid entropy states are calculated by numerically solving Eqs. (6) and (7). Since the vorticity in the background is given by Eq. (8), Eq. (6) is a 2-D nonlinear partial differential equation. Parameters β, Ω, μ, and $\{\mathbf{R}_i\}$ are searched to obtain the maximum fluid entropy state for given conserved quantities H, L, Γ, as well as the M vorticity distributions $\omega_{v,i}$ of the strong vortices. In general, there may be several possible patterns of the strong vortices, representing local maxima of the entropy. To obtain maximum entropy states close to the experimental vortex crystals, we use the experimental positions of the strong vortices as the initial values of the search. The procedure we employed to determine M and $\omega_{v,i}$ from the experimental data is essentially that of McWilliams [9].

The complete specification of the problem requires the choice of a value for the Fermi vorticity ω_f associated with the microscopic background vorticity elements. Although in general ω_f can have a distribution of vorticity levels [8], a single value is found to be sufficient to explain the experiments. The maximum of the measured vorticity is taken as ω_f. This choice is based on the following considerations: ω_f must be larger than or equal to the maximum observed background vorticity level, which is coarse-grained by the experimental imaging system; since the observed vortex

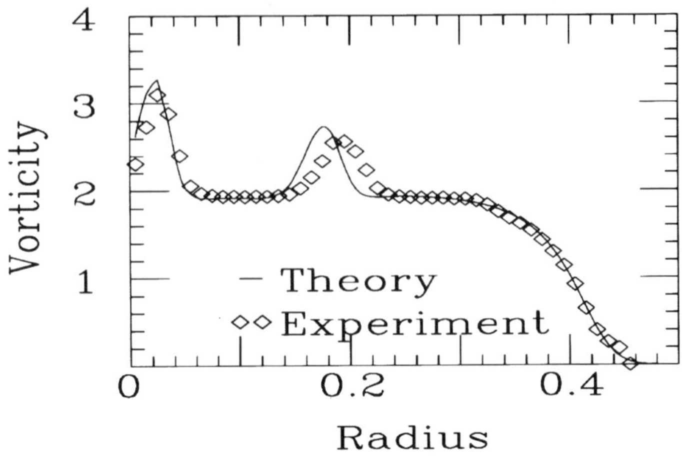

FIGURE 4. The θ-averaged vorticity profile from the five vortex state of FIG. 3(c).

crystal states appear to be nearly Fermi-degenerate, in some finite region of the background the coarse-grained vorticity should approach ω_f. The following fundamental units are used to renormalize the physical quantities: length is r_w (radius of the circular conducting boundary) and vorticity is Γ / r_w^2.

The theory agrees well with the experimental data. The calculation reproduces the observed vortex crystal patterns, as shown in FIG. 3, in which selected experimental images of vortex crystals and their corresponding maximum fluid entropy states are displayed. Although there is a slight azimuthal asymmetry and some noise in some experimental images, the observed background vorticity is close to the theory, as shown in the θ-averaged vorticity profiles in FIG. 4. The background of the finite temperature solutions has the following features: The edge falls off over a scale related to the fluctuation energy $1/\beta$ of the vorticity elements near the edge, as in FIG. 4; and near a strong vortex the background vorticity is slightly depressed, since ϕ tends to increase due to the influence of the strong vortex, as observed around the large central vortex in FIG. 3(d).

4. DISCUSSION

The maximum fluid entropy theory is successful in describing the vortex crystal states. This provides us with an intuitive picture of vortex crystal formation after strong vortices have formed: the turbulent evolution ergodically mixes certain regions of the flow (the background) and this process causes the unmixed regions of strong vorticity to relax to an observed equilibrium pattern (the crystal). The increase in the order of the strong vortices during this process is offset by the increased disorder in the background, as expected from the second law of thermodynamics.

However, maximum fluid entropy theory does not answer many interesting questions related to the dynamics of vortex crystal formation. For example, the details of the mixing process and the rate at which strong vortices relax to the equilibrium patterns remain to be investigated. Also, experimentally, vortex crystals are not formed from some initial conditions that are similar to those that do lead to the formation of vortex crystals [5]. The characteristics of the initial conditions for vortex crystals is still unknown.

REFERENCES

1. McWilliams, J.C. 1983. On the relevance of two-dimensional turbulence to geophysical fluid motions. J. Méc. Theor. Appl., Spec. Suppl. 83–97.
2. Ingersoll, A.P. 1990. Atmospheric dynamics of the outer planets. Science. **248:** 308–315.
3. Marcus, P.S. 1988. Numerical simulation of Jupiter's Great Red Spot. Nature **331:** 693–696.
4. Driscoll, C.F. & K.S. Fine. 1990. Experiments on vortex dynamics in pure electron plasmas. Phys. Fluids. **2:** 1359–1366.
5. Fine, K.S., A.C. Cass, W.G. Flynn & C.F. Driscoll. 1995. Relaxation of 2D turbulence to vortex crystals. Phys. Rev. Lett. **75:** 3277.
6. Jin, D.Z. & D.H.E. Dubin. Phys. Rev. Lett. Accepted for publication.
7. Carnevale, G.F. *et al.* 1991. Evolution of vortex statistics in two-dimensional turbulence. Phys. Rev. Lett. **66:** 2735–2737.
8. Miller, J. 1990. Statistical mechanics of Euler equations in two dimensions. Phys. Rev. Lett. **65:** 2137–2140; Robert, R. & J. Sommeria. 1991. Statistical equilibrium states for two-dimensional flows. J. Fluid Mech. **229:** 291–310.
9. McWilliams, J.C. 1990. The vortices of two-dimensional turbulence. J. Fluid Mech. **219:** 361–385.

Simulations of Cellular Automata

HARMEN J. BUSSEMAKER[a]

Department of Physics, University of Maryland, College Park, Maryland 20742

Fluid-type lattice gas automata (LGA) have provided a fruitful method for studying spatially extended systems of interacting particles. LGA are discrete dynamical systems in which the space variable is restricted to a lattice and the time variable is integer-valued. In addition, the velocity states are discrete and finite in number. The dynamics of LGA is specified by free motion for each particle between lattice points on each time step, followed by "local collision rules" specifying transitions to a new set of velocity states depending on the particles at each node. Fluid-type LGA are further restricted to have collision rules respecting mass and momentum conservation, and a lattice structure consistent with macroscopic fluid dynamics at large space and time scales. A primary advantage of such models is that they are ideally suited to the architecture of modern computers and the consequent efficient simulation of the dynamics using massively parallel computer algorithms. Originally motivated as an alternative means to study nonlinear fluid dynamics (e.g., turbulence), LGA have proved to be fruitful testing grounds for the methods of nonequilibrium statistical mechanics. In this presentation we describe a recent study of long-range correlations and pattern formation using LGA. The adaptation of kinetic theory methods to LGA allow more detailed comparisons with computer simulations than are possible for real fluids.

LGA with strictly local collision rules that violate detailed balance have a non-Gibbsian equilibrium state, and we have studied equal time spatial correlations in this state. The existence of local conservation laws for mass and momentum leads to the existence of long-range algebraic decay of the pair correlation function. The primary theoretical analysis is based on a ring kinetic theory going beyond the Boltzmann equation for the occupation numbers to describe equal time correlations as well [1], [2]. The kinetic equations have been solved numerically to determine both the equilibrium occupation numbers and the correlation functions for an initial state which approaches the asymptotic equilibrium state. In this way it is possible to study both the equilibrium correlations and the approach to equilibrium. One important result is the confirmation of the mode coupling mechanism for long-range correlations as well as long time algebraic approach to equilibrium [3]. Two specific LGA without detailed balance collision rules were considered, a system of interacting random walkers on a square lattice with Fermi exclusion rules, and a fluid-type LGA defined on a triangular lattice. The agreement between computer simulation of the microdynamics and the results of kinestic theory are excellent.

[a]Present address: The Rockefeller University, Box 25, 1230 York Avenue, New York, New York 10021.

REFERENCES

1. ERNST, M.H. & H.J. BUSSEMAKER. 1995. J. Stat. Phys. **81:** 515–528.
2. BUSSEMAKER, H.J. & M.H. ERNST. 1996. Phys. Rev. E. **53:** 5837–5851.
3. BRITO, R., H.J. BUSSEMAKER, M.H. ERNST & J. MATSUI. 1995. Phys. Rev. E. **52:** 2657–2667.

Collisionless Relaxation in Galactic Dynamics and the Evolution of Long-Range Order

HENRY E. KANDRUP[a]

Department of Astronomy and Department of Physics and Institute for Fundamental Theory, University of Florida, Gainesville, Florida 32611

ABSTRACT: This paper presents a critical assessment of certain aspects of collisionless galactic dynamics, focusing on the interpretation and limitations of the collisionless Boltzmann equation and the physical mechanisms associated with collisionless, or near-collisionless, relaxation. Numerical and theoretical arguments are presented to motivate the idea that the evolution of a system far from equilibrium should be interpreted as involving nonlinear gravitational Landau damping, a picture which implies a greater overall coherence and remembrance of initial conditions than is implicit in the conventional paradigm of violent relaxation.

INTRODUCTION AND MOTIVATION

The principal aim of the work described here is to understand the gravitational N-body problem for a collection of nearly point "particles" of comparable mass, such as, for example, the stars in a galaxy. This N-body problem is not directly applicable for problems involving galaxy formation, or in understanding the structure and evolution of spiral galaxies, where dissipative gas dynamics must play an important role. However, it should be relevant in understanding the structure of elliptical galaxies, gas-poor [1] objects which are typically assumed to be in or near a (meta-) equilibrium [2]. Moreover, it should be important in understanding the evolution of elliptical galaxies when, as must often occur [3], they are displaced from their near-equilbrium by a collision or other close encounter with some companion object.

The conventional wisdom of galactic dynamics [2] asserts that, over sufficiently short time scales, the structure and evolution of a collection of N self-gravitating point masses can be described by the *collisionless Boltzmann equation*, the gravitational analog of the Vlasov–Poisson system of plasma physics. Alternatively, it is assumed that, on longer time scales, one must allow for discreteness effects, described by the *collisional Boltzmann*, or *Fokker–Planck equation*, the latter being simply the gravitational analog of the Landau equation from plasma physics.

In this context, it is typically assumed (cf. [4]) that there are four distinct sorts of "relaxation" processes, namely, (1) "violent relaxation," (2) gravitational Landau damping, (3) phase mixing, and (4) "collisional relaxation." Violent relaxation [2], [5] refers to the early stages of evolution in which a system, initially far from any

[a]Supported in part by National Science Foundation Grant No. PHY92-03333. Some of the numerical calculations were facilitated by computer time provided by IBM through the Northeast Regional Data Center (Florida). This article is based in part on collaborations with Elaine Mahon, Christos Siopis, and Barbara Eckstein, students at the University of Florida who have been supported by NASA through the Florida Space Grant Consortium.

steady state, evolves towards some meta-equilibrium, i.e., a time-independent solution to the collisionless Boltzmann equation (CBE). By contrast, Landau damping and phase mixing reflect smaller amplitude effects acting in a system which is much closer to a meta-equilibrium. Collisional relaxation is associated with discreteness effects which, dating back to Chandrasekhar [6], have typically been modeled as an incoherent sum of binary encounters, i.e., gravitational Rutherford scattering.

The first three of these mechanisms, incorporated in the CBE, are presumed to induce an efficient approach towards some meta-equilibrium on a time scale of order t_{cr}, a characteristic crossing time for the system in question. By contrast, collisional relaxation is presumed to be important only on a much longer relaxation time t_R which, for systems as large as an entire galaxy, is orders of magnitude longer than t_H, the age of the Universe. In the language of statistical physics, one thus assumes that a mean field theory, based on the CBE, accurately describes the structure and evolution of a gas-poor elliptical galaxy over time scales shorter than or comparable to the age of the Universe.

In this paper, I play devil's advocate and ask to what extent one can really be confident that this conventional wisdom is completely correct. Specifically, I address four distinct questions: (1) Does one really know that the CBE is the whole story? (2) In what sense, and to what extent, does the CBE imply an evolution towards some meta-equilibrium? (3) How should one interpret the flow associated with the CBE and, related to this, what is the dynamical principle that determines the form of the meta-equilibrium towards which an N-body system is presumed to evolve? (4) What precisely is the connection between the CBE and the N-body problem, and, in particular, what do CBE characteristics computed for a smooth initial distribution function have to do with orbits generated as a solution to the full gravitational N-body problem?

The principal substantive conclusion is that, contrary to what is usually assumed by galactic dynamicists, phase mixing, Landau damping, and violent relaxation should be viewed as (not so different) manifestations of a single phenomenon. As is well known to plasma physicists, linear Landau damping *is* in fact a type of phase mixing. Moreover, numerical experiments and theoretical considerations suggest that the evolution of a system far from equilibrium can be interpreted fruitfully as entailing nonlinear Landau damping, this implying a greater overall coherence and remembrance of initial conditions than is implicit in the conventional picture of violent relaxation.[1]

IS THE COLLISIONLESS BOLTZMANN EQUATION THE WHOLE STORY?

The basic assumption underlying the CBE is that the system in question can be described probabilistically in terms of a smooth one-particle distribution function,

[1]It is ironic that Lynden-Bell's original paper [5] on violent relaxation does not implement clear cut distinctions amongst violent relaxation, Landau damping, and phase mixing. Rather, it suggests a more integrated approach which views phase mixing-cum-Landau damping as an important part of violent relaxation, an approach more consistent, in some respects, with the ideas advocated in this paper than with conventional wisdom.

$f(x^a, v_a, t)$, which can be interpreted as a phase space number or mass density that yields the probability of finding a particle near the spatial point x^a with velocity v_a at time t. (In what follows, it will be assumed that each particle has a mass $m = 1$ since, in the context of a mean field theory, allowing for a distribution of particle masses has absolutely no effect.) In the context of the initial value problem, this implies that the true initial distribution function, given as a sum of N phase space delta functions, has been replaced by a smooth function $f(x^a, v_a, 0)$.

The CBE manifests the idea that the distribution function f will free stream in the self-consistent potential Φ associated with f, so that

$$\frac{df}{dt} \equiv \frac{\partial f}{\partial t} + v^a \frac{\partial f}{\partial x^a} - \frac{\partial \Phi}{\partial x^a} \frac{\partial f}{\partial v_a} = 0\,, \tag{1}$$

where

$$\nabla^2 \Phi(x^a, t) = 4\pi G \int d^3 v f(x^a, v_a, t). \tag{2}$$

In the context of a many-particle phase space description, this is equivalent to assuming that the two-particle distribution function $g(1, 2)$ factorizes into a product of one-particle f's, so that $g(1, 2) \approx f(1) f(2)$, i.e., entirely neglecting the effects of particle-particle correlations. It is assumed that this neglect is justified on time scales short compared with the natural time scale t_R associated with gravitational Rutherford scattering, a time scale which, for very large N, is much larger than t_{cr}.

Unfortunately, this neglect of particle correlations constitutes an uncontrolled approximation. Thus, for example, unlike the case of a neutral plasma, one cannot derive the next-order correction to the CBE in the context of some systematic perturbation expansion. Physically, the problem is that there is no shielding to vitiate the long-range $1/r^2$ force. This is, for example, manifested by the fact that a $1/r$ potential yields an infinite cross section, so that, when evaluating the effects of binary encounters in the usual way for an infinite and homogeneous system, one encounters logarithmic divergences in the limit of large impact parameter (the problematic Coulomb logarithm of galactic dynamics).

Physically, one might hope to circumvent this difficulty by first identifying the bulk mean field force acting at any given point in space and then treating "fluctuations away from mean field conditions" as the source of deviations from mean field theory. For $N \gg 1$ one might expect that these fluctuations are small, so that their effects do in fact constitute a small perturbation. This splitting into mean field plus fluctuations can be introduced formally, for example, by using time-dependent projection operators [7], but is difficult to implement concretely because of the apparent absence of a clean separation of time scales.

One case where one can try to perform a concrete calculation is for the toy model of a system that is rotating with an angular velocity Ω carefully chosen to ensure that the mean field equilibrium is homogeneous (here the rotation provides a fictitious force that plays the role of the uniform oppositely charged background in a plasma). In this case, one finds that the screened potential $\exp(-\kappa r)/r$ of plasma physics is replaced by an oscillatory potential $\exp(i\kappa r)/r$, this reflecting the fact that an homogeneous system is unstable towards sufficiently long wavelength perturbations (the Jeans instability). The conventional response is to introduce by hand a cutoff at a length scale $R_{\rm sys}$ comparable to the size of the system, supposing that the system was generated at some earlier time by the fragmentation of a much larger, nearly homogeneous entity that experienced the Jeans instability.

As demonstrated by Thirring and coworkers (cf. [8]), there *does* in fact exist a well defined thermodynamic limit in which gravitational mean field theory becomes exact. However, this limit is an unusual one. Given that realistic self-gravitating systems are inhomogeneous, one cannot simply introduce the ordinary thermodynamic limit, with number $N \rightarrow \infty$ and volume $V \rightarrow \infty$ but finite density $n = N/V$. Instead, one must allow for a limit in which the coupling constant G also scales. The problem, however, is that there is no guarantee that this limit is physically realistic and that, even assuming that it is physical, there are no rigorous (or even quasi-rigorous) estimates as to the time scale $t_R(N)$ on which the mean field description would be expected to fail.

Given that theoretical analyses have as yet proven inconclusive, one might instead seek recourse to numerical experiments. This, however, is difficult. For generic systems not characterized by a high degree of symmetry, the CBE is a partial differential equation in six independent phase space variables which is extremely expensive to solve computationally. Moreover, the time required to solve the N-body problem using an honest direct summation code scales as N^2, so that the consideration of very large N, where the CBE is expected to be valid, becomes prohibitive. The approximate validity of the CBE has been corroborated in the sense that N-body realizations of meta-equilibria often appear to behave stably for relatively short time scales, and that, at least for short time scales, various velocity moments extracted from N-body simulations and time-dependent solutions to the CBE yield no gross discrepancies. However, computational limitations have prevented truly detailed comparisons of N-body and CBE evolutions hitherto.

There is in fact one concrete setting where detailed computations have been done, namely the toy model of one-dimensional gravity, i.e., a collection of infinite plane sheets characterized by a two-body potential [9]

$$V(|x_i - x_j|) = G|x_i - x_j|. \tag{3}$$

For this simple model, the phase space is only two-dimensional and the forces are trivial computationally, so that it is relatively easy to perform highly accurate computations. The net result of such simulations is that, at least for times as long as 10^2–$10^3 t_{cr}$, there is a reasonably good overall agreement between N-body simulations and an evolution governed by the CBE [10]. Interestingly, however, the evolution seems to be substantially more complex than what most astronomers would expect for three-dimensional gravity (cf. [11], [12]). These conclusions seem highly suggestive. However, they must both be taken with a fair grain of salt since there is no compelling reason to believe that the evolution of one-dimensional gravitational systems is qualitatively similar to the evolution of "real" three-dimensional systems.

In summary, even though a mean field theory based on the CBE may seem well motivated physically, there is as yet no rigorous proof of its validity and, in particular, no rigorous estimate as to the time scale on which it might be expected to fail. Moreover, as discussed below, there are reasons to suspect that even very small corrections could be important by accelerating the approach towards equilibrium and, especially, by serving to "fuzz out" small scale structures predicted by the CBE to arise as a self-gravitating system approaches some meta-equilibrium.

IN WHAT SENSE DOES THE COLLISIONLESS BOLTZMANN EQUATION IMPLY AN APPROACH TOWARDS EQUILIBRIUM?

In addressing the question of evolution, the most important thing to recognize about the CBE is that it, like the true N-body problem, is Hamiltonian. Specifically, the CBE constitutes a noncanonical Hamiltonian system, formulated in an infinite-dimensional phase space, where the fundamental dynamical variable is the distribution function f itself [13], [14]. The proof that the CBE is Hamiltonian entails the identification of a Hamiltonian functional $\mathcal{H}[f]$ and a Lie bracket [., .] defined on functionals $\mathcal{A}[f]$ and $\mathcal{B}[f]$, so chosen that the CBE takes the form

$$\frac{\partial f}{\partial t} + [\mathcal{H}, f] = 0\,. \tag{4}$$

It is straightforward to verify that the antisymmetric operation

$$[\mathcal{A},\mathcal{B}] = \int d^3x d^3v f\left\{\frac{\delta\mathcal{A}}{\delta f}, \frac{\delta\mathcal{B}}{\delta f}\right\}, \tag{5}$$

with $\delta/\delta f$ a functional derivative and {., .} the ordinary Poisson bracket of particle mechanics, defines a *bona fide* Lie bracket acting in the infinite-dimensional phase space of distribution functions. However, if one identifies $\mathcal{H}[f]$ as the mean field energy, i.e.,

$$\mathcal{H}[f] =$$

$$\frac{1}{2}\int d^3x d^3v v^2 f(x^a, v_a, t) - \frac{G}{2}\int d^3x d^3v \int d^2x' d^3v' \frac{f(x^a, v_a, t) f(x'^a, v_a', t)}{|\mathbf{x}-\mathbf{x}'|}, \tag{6}$$

it is easy to see that, with this bracket, Eq. (4) reduces to the CBE in the form

$$\frac{\partial f}{\partial t} - \{E, f\} = 0\,, \tag{7}$$

where $E = \frac{1}{2}v^2 + \Phi[f]$ is the particle energy.

Associated with this Hamiltonian character is the existence of an infinite number of conserved quantities, the so-called Casimirs, which generalize the notion of conservation of phase (or Liouville's Theorem) for a distribution function f evolving in an external potential [15]. Specificially, one knows that, for any function $\chi(f)$, the numerical value of $C[f] = \int d^3x d^3v \chi(f)$ is independent of time.[2] One concrete manifestation of these Casimirs is the fact that any phase-preserving perturbation of a CBE equilibrium must be generated by a canonical transformation [16]. Every time-independent equilibrium solution f_0 corresponds to an extremal point of the mean field energy, i.e., $\delta^{(1)}\mathcal{H} \equiv 0$ for all phase preserving perturbations of the equilibrium.

The Hamiltonian character of the evolution is crucial because it precludes entirely the possibility of any pointwise approach towards a time-independent equilibrium. Only by replacing the true distribution function f by some coarse-grained $\bar{f}$ can one hope to identify an object which actually approaches equilibrium. Any meaningful discussion of "the approach towards equilibrium" must be formulated in the context of a coarse-grained distribution function or, alternatively, some set of coarse-grained observables constructed from the true distribution function by some phase space averaging.

Very little is known mathematically about an evolution governed by the CBE. Recently, however, it has been proven that the CBE manifests global existence [17], [18]. Specifically, one knows that, given sufficiently smooth initial data (C^{1+}), the distribution function will never diverge. In other words, unlike the case of a perfect fluid, the evolution never leads to caustics or shocks. This suggests in particular that smooth initial conditions cannot evolve near-singular cores or other lumpy exotica of the form discovered recently using the Hubble Space Telescope [19].

Unfortunately, however, there are no rigorous results about a (suitably coarse-grained) approach towards equilibrium. Rather, the only results obtained to date relate to the asymptotic behavior of time integrals of quantities like $\mathcal{K}[f]$ or $\mathcal{V}[f]$, the mean field kinetic and potential energies. Thus, for example, in certain cases one can prove [20] the existence of quantities like

$$K \equiv \lim_{T \to \infty} \frac{1}{T} \int_0^\infty \mathcal{K}[f(t)]dt \,. \tag{8}$$

However it has not been possible to prove statements about the behavior of (say) $\mathcal{K}[\bar{f}]$ at a fixed instant of time, let alone the behavior of the full $\bar{f}$.

One might naively suppose that this simply reflects the fact that any such result would be very hard to prove. However this is not completely true. Specifically, there exist exact, time-dependent solutions to the CBE, corresponding to a system that remains bounded in space, that exhibit no approach towards a coarse-grained equilibrium [21], [22]. Rather, these solution correspond to finite amplitude, undamped oscillations about an otherwise time-independent equilibrium f_0.

There is of course no reason to expect *a priori* that such oscillating solutions actually exist in nature. However, it appears that they can arise naturally at least in the context of one-dimensional gravity [10]. Specifically, for both N-body simulations and the CBE, the evolution of counter-streaming initial conditions (two equilibria engaged in a head-on collision) can yield a final $f(t)$ which involves undamped oscillations about a time-independent f_0 which, significantly, contains "holes," i.e., regions in the middle of the occupied phase space region where $f_0 \to 0$.

[2]Significantly, however, even though the CBE admits this infinite collection of conserved quantities, it is extremely unlikely that it is integrable (unlike, for example, the Korteweg–de Vries equation).

HOW SHOULD ONE INTERPRET THE FLOW ASSOCIATED WITH THE COLLISIONLESS BOLTZMANN EQUATION?

"Violent Relaxation"

Most discussions of the evolution of a self-gravitating system initially far from a meta-equilibrium have been formulated within the context of the paradigm of violent relaxation, a sort of "self-consistent egg beater" model originally proposed by Lynden-Bell thirty years ago [5]. Different workers have interpreted this paradigm in rather different ways. Indeed, the conventional wisdom would appear to ignore many of the fine points made in the original paper. However, (almost) all versions of violent relaxation seem to incorporate four basic ingredients:

1. Attention focuses on a coarse-grained distribution function $\overline{f}$ which can "fuzz out" in a way that is impossible for the true f, which is strongly constrained by conservation of phase. The obvious point here is that if f satisfies the CBE, in general the coarse-grained $\overline{f}$ will not. Lynden-Bell starts from the assumption that the fundamental object of interest is a smooth one-particle f that satisfies exactly the CBE and then implements a coarse-graining of the one-particle phase space to extract an appropriate $\overline{f}$. However, many later workers have sought to relate $\overline{f}$ directly to the full N-particle dynamics, as derived, e.g., from numerical simulations. This difference in perspective is important. Lynden-Bell assumes that the CBE is exact in some suitable $N \to \infty$ continuum limit and formulates his entire theory in that limit. For him, therefore, violent relaxation says little if anything about individual particles, and should be interpreted instead as tracking a coarse-grained free streaming of phase space fluid elements.

2. One invokes the effects of a strongly time-dependent bulk potential $\Phi[f]$ to help "shuffle" the coarse-grained $\overline{f}$. The initial $f(0)$ yields the mass density $\rho(0)$ and, hence, the gravitational potential $\Phi(0)$ which determines the initial motion of each phase point in the system. In general, $f(0)$ will be far from equilibrium, so that Φ will induce changes in f on a time scale $\sim t_{cr}$. However, changes in f correspond in general to changes in Φ, so that the potential itself will vary significantly on the same time scale t_{cr}. This implies in turn that the particle energy $E = \frac{1}{2}v^2 + \Phi$ associated with any characteristic will be strongly time-dependent. In particular, if the potential exhibits large enough variations on a sufficiently short time scale, E will be far from an adiabatic invariant, so that one might expect that the energies of the different characteristics will become "well shuffled" and/or randomized.

3. One supposes that, as a result of this convulsive process, $\overline{f}$ will evolve towards, albeit not necessarily to, the "most likely" $\overline{f}_0$ consistent with an appropriate set of constraints. Two different types of constraint are usually considered, namely (1) bulk holonomic constraints, such as conservation of energy and particle number, and (2) the residual, coarse-grained effects of conservation of phase, which may still prove important if the system is sufficiently dense. Although crucial conceptually, the latter are often unimportant as a practical matter (unless, for example, the system is very dense) and, for

that reason, will be ignored in the following.

4. One supposes further that the "most likely" $\overline{f}_0$ can be derived from a combinatoric argument by determining the macrostate corresponding to the largest number of microstates. To the extent that the constraints associated with conservation of phase are unimportant, this leads to the identification of the most likely $\overline{f}_0$ as being that particular coarse-grained distribution function that maximizes the Boltzmann entropy:

$$S[\overline{f}] = -\int d^3x \int d^3v \, \overline{f} \log \overline{f}. \tag{9}$$

It is easily seen that, by combining these four ingredients, one is led to a predicted evolution towards an isothermal distribution of the form

$$\overline{f}_0 \propto \exp\{-\beta(v^2/2 + \Phi[\overline{f}_0])\}. \tag{10}$$

(Allowing for coarse-grained conservation of phase would modify Eq. (9) to yield a degenerate isothermal.)

Lynden-Bell would stress that, albeit involving combinatoric arguments familiar from information theory, his derivation of Eq. (10), which deals with phase space fluid elements rather than discrete particles, is not an entropy argument in the usual sense. One clear indication of this is the fact that the particle mass m does not enter into $\overline{f}_0$. If one implements a "true" entropy argument for a collection of particles with a distribution of masses, instead one finds [64] a mass-dependent extremal entropy configuration of the form $\overline{f}_0 \propto \exp\{-\beta(mv^2/2 + m\Phi[\overline{f}_0])\}$.

The "prediction" of an isothermal was considered a triumph back in the 1960's, when the observed luminosity of many, if not all, elliptical galaxies could be well fit by truncated isothermals, i.e., the so-called King models [23]. However, in the intervening years, improved high resolution photometry and better background subtraction have yielded better data for which truncated isothermals do not provide adequate fits. Instead, one is led typically to model data using a de Vaucouleur [24] density profile or, more recently, a Nuker density law that allows for a central cusp [25].

Even neglecting the fact that observed galaxies are not well fit by truncated isothermals, the preceding doctrine of violent relaxation is open to several criticisms. Perhaps the most obvious is simply: why should one assume that "most likely" means (essentially) "maximize the Boltzmann entropy"? The usual justification for this standard thermodynamic assumption, and for the combinatoric arguments exploited by Lynden-Bell, simply fails for a system interacting via a long-range force like gravity. The standard argument relies crucially on the assumption of approximate extensivity, i.e., the assumption that, if one's system be viewed as a sum of individual pieces, quantities like the total energy are well approximated by a sum of the corresponding quantities for the individual pieces [26]. Indeed, one could in principle play the same maximization game for any convex phase space functional [27], which leads to a natural connection with Casimir arguments from plasma physics [28].

More pragmatically, one infers from N-body simulations that a strongly convulsing mean field potential is not necessary. One observes a comparably efficient approach towards a meta-equilibrium on a time scale $\sim t_{cr}$ both for "violent" evolution, where Φ exhibits huge changes on very short time scales, and for "nonviolent" evolution, where Φ exhibits only relatively small changes [29]. Nonviolent relaxation can be just as efficient as violent relaxation.

Perhaps the most striking feature about violent relaxation is the assumption that the evolution of $\overline{f}$ is completely unconstrained aside from the imposition of some small number of bulk holonomic constraints and coarse-grained conservation of phase.[3] Seemingly, this implies that when viewed in terms of $\overline{f}$, the flow is chaotic, modulo only a restriction to a suitably defined phase space hypersurface on which bulk quantities like the total energy are constant.

That this assumption is suspect was already recognized when violent relaxation was first proposed [30], but at that time it was not possible to implement direct tests computationally. However, during the past fifteen years others (cf. [31]–[33]) have exploited improved computational resources to test this assumption, and have been led to the conclusion that it is *not* justified. Specifically, analyses of N-body simulations have demonstrated that there is a precise sense in which individual particles tend to "remember" certain aspects of their initial conditions, at least statistically. Thus, for example, for a variety of different geometries, allowing for both collapse simulations and collisions between galaxies, one discovers that particles that start with small binding energies tend to end up with small binding energies, and visa versa. In the context of the CBE, this would seem to indicate that phase space cells with different initial energies do not become "well shuffled."

This effect can be quantified both microscopically and mesoscopically. For example, one can order the particles in a simulation in terms of their binding energies at different times and then compute quantities like the rank correlation $\mathcal{R}(0, t)$ between the initial ordering and the ordering at some later time t [33]. The net result of such investigations is that, even allowing for an extremely violent evolution, where the virial ratio $2\mathcal{K}/\mathcal{V}$ changes by factors of two or more within a crossing time t_{cr}, $\mathcal{R}(0, t) > 0.6$ at $t = 20t_{cr}$, well after the system appears to have settled down towards a meta-equilibrium.

Alternatively, one can bin the particles in a simulation in terms of their initial binding energies, and then study the properties of the different bins as the system evolves into the future. Here the obvious question is whether, and if so to what extent, the statistical properties of particles in the initial low and high energy bins converge, as would be expected if the particle energies become completely randomized [32], [33]. The conclusion of such an analysis is that, at least over time scales $\ll t_R$, the mean energies of the different bins show absolutely no tendency to converge, although the energy dispersions of the different bins do grow.

The seemingly unambiguous conclusion derived from these numerical investigations is that, contrary to what is usually assumed in violent relaxation, the evolution towards a meta-equilibrium can be comparatively nonviolent, and that, even for a violent evolution, particle energies are not totally randomized. One cannot assume a completely shuffled distribution of particle energies. There is thus a need for an al-

[3]As Lynden-Bell discussed in an appendix to his paper, it is straightforward to generalize his derivation of Eq. (10), which only conserved the mean field energy (6), to allow for conservation of the mean field angular momentum or any other quantity that can be written as a phase space integral.

ternative theory of "nonviolent relaxation" [32], [34] which does not assume a completely chaotic flow.

Alternatives to Violent Relaxation

When considering a plasma which is displaced only slightly from some meta-equilibrium f_0, it is customary to visualize evolution in terms of linear Landau damping [35] which, for simple geometries, implies that perturbations in the density and potential, $\delta\rho$ and $\delta\Phi$, damp exponentially. Physically, this damping is interpreted (cf. [36]) as reflecting a resonant coupling of (unperturbed) particles with physical velocity v^a and a wave (the perturbation) with phase velocity c^a. Note that this picture necessarily involves a coarse-grained description since it is formulated in terms of configuration space observables like ρ and Φ rather than the full distribution function f.

It is difficult to interpret, let alone implement, Landau's original calculations for systems that are strongly inhomogeneous. However, there is an alternative interpretation which continues to make sense, namely that Landau damping involves the phase mixing of a superposition of normal modes. In particular, the distinction which galactic dynamicists are wont to make (cf. [2]) between phase mixing and Landau damping is simply not valid. Viewed appropriately, *linear Landau damping is a type of phase mixing.*

A small perturbation δf governed by the Vlasov equation appropriate for a neutral electrostatic plasma will satisfy a linearized evolution equation which can be written symbolically in the form

$$\frac{\partial \delta f}{\partial t} = \mathcal{A}\delta f, \tag{11}$$

where $\mathcal{A}$ is a linear integro-differential operator. Suppose, however, that the initial perturbation $\delta f(0)$ is expanded in terms of the normal modes of $\mathcal{A}$. If the spectrum of $\mathcal{A}$ is continuous, a smooth initial perturbation necessarily involves a superposition of normal modes, but the evolution of this initial perturbation then implies that quantities like $\delta\rho$ and $\delta\Phi$ necessarily damp exponentially [37]. In other words, Landau damping is guaranteed. This is intrinsically a phase mixing process, involving the nondissipative spreading of an initial wavepacket. The perturbation manifests an initial coherence that is lost as it evolves into the future. The full δf itself does not tend towards zero in any smooth, pointwise sense, but coarse-grained observables like $\delta\rho$ and $\delta\Phi$ do.

Alternatively, the existence of one or more discrete modes implies that there are perturbations which do not phase mix or Landau damp. The classic example thereof is van Kampen [38] modes, which correspond to waves executing undamped oscillations with a phase velocity c^a where resonant coupling to the unperturbed particles is impossible because $f_0(c^a) \equiv 0$.

The important point here is that whether or not Landau damping/phase mixing always occurs is determined completely by whether or not the spectrum of $\mathcal{A}$ is purely continuous, and that the specific form of the modes is irrelevant. For this reason, it is possible to prove an analogous result for small perturbations of a gravitational CBE equilibrium [39]. Specifically, if the modes of the linearized perturbation equation are all continuous, any linearized perturbation will always exhibit Landau damping/phase mixing. If, however, discrete modes exist, there are possible initial perturbations which do not damp.

Determining whether or not the normal modes for realistic CBE equilibria are continuous is extremely difficult. In general it seems impossible to calculate the modes explicitly, and a formal analysis is also hard since $\mathcal{A}$ is not an elliptic operator and evolves a singular integral kernel. It is, however, worth recalling that, at least in a plasma, discrete modes are typically associated with nontrivial boundary conditions, e.g., spatial confinement or the existence of a maximum velocity. Thus, in particular, exponential damping is guaranteed if f_0 is analytic. However, one can argue that self-gravitating CBE equilibria always have nontrivial boundary conditions. They are always confined to a compact spatial region by the self-consistently determined potential; and there always exists a maximum speed $|v_{max}(x^a)|$ since characteristics for which v is too large will quickly escape to infinity.

For this reason, one might reasonably conjecture [39] that many, but not necessarily all, perturbations will Landau damp. Most initial $\delta\rho$'s might be expected to damp, but undamped perturbations could exist for modes that manifest coherence on a scale comparable to the size R_{sys} of the system, or, especially, given nontrivial phase space structures. In this regard, it seems significant that the aforementioned models of galaxies that exhibit undamped oscillations [21], [22] all contain phase space "holes," i.e., regions in the middle of the occupied phase space regions where $f_0 \to 0$.

The discussion hitherto has focused on linear Landau damping, i.e., the behavior of a small initial perturbation which satisfies a linearized evolution equation. However, at least in principle this picture can also be extended to finite amplitude perturbations. In the context of plasma physics, this means considering nonlinear Landau damping, which generalizes the linear theory by incorporating the possibility of mode-mode couplings that allow energy transfer between different modes. Because it is very difficult in a realistic gravitational setting to determine the form of the normal modes even for a linearized perturbation, the development of a detailed theory of nonlinear gravitational Landau damping would be extremely difficult. However, despite this problem it is still possible to visualize self-gravitating systems in an identical fashion, even though the forms of the modes are not known explicitly.

It would thus seem reasonable to *ask whether it is possible, and fruitful, to interpret the evolution of a generic initial f(0) towards some meta-equilibrium as a manifestation of nonlinear Landau damping.* Such a picture, based on a nonlinear evolution driven by mode-mode coupling, is by construction a relatively coherent process, consistent with the statistical regularities observed in N-body simulations and significantly less "incoherent" than what is implicit in the standard picture of violent relaxation.

Is there any evidence that this interpretation is actually viable? One reassuring point is that, at least naively, one might expect the time scales to work out correctly. Indeed, by analogy with Landau damping in a plasma, as well as from simple dimensional analysis, one would expect that a perturbation manifesting coherence on a scale $\sim R_{sys}$ should damp on a time scale of order t_{cr}. Strictly speaking, this dimensional argument is only justified for the linear theory, since mode-mode couplings could introduce new time scales related to t_{cr} by ratios of different length scales. However, at least for plasmas one can show that in certain cases, allowing for nonlinear mode-mode couplings does not lead to new time scales [40].

The intuition that dynamical evolution can lead to undamped oscillating modes is of course corroborated by the aforementioned one-dimensional simulations of both gravitational and plasma systems with counter-streaming initial conditions [10]. Here the late time state involves reasonably large amplitude oscillations, corresponding to nearly periodic variations in the virial ratio $2\mathcal{K}/\mathcal{V}$ at the 10% level, characterized by a period $\sim t_{cr}$ and a coherence length $\sim R_{sys}$. Unlike the exact pulsating solutions [21], [22], which incorporate true periodicity, the late time state arising in these one-dimensional simulations is not exactly periodic. Rather, what one finds is that the nonoscillating, nearly time-independent, component f_0 slowly changes as, via phase mixing ("phase wrapping"), it manifests a coarse-grained approach towards a truly time-independent f_0.

Another point, not conclusive but arguably suggestive, arises from a consideration of moment codes. These codes approximate the Vlasov equation of plasma physics or the CBE of gravity by a discrete collection of coupled moment equations which have been reduced to a finite set through the introduction of a symplectic trunction (for example, Channell's [41] BEDLAM code for an electrostatic plasma and the corresponding GADFLI for a self-gravitating system each involve a collection of 203 coupled moments). When used by accelerator dynamicists to study the stability of beam flows, these efficient moment codes have proven spectacularly successful in the sense that they agree with theory and N-body simulations (and experiments!), and one might expect analogously that their gravitational analogs could prove powerful tools in diagnosing stability or lack thereof for solutions to the CBE.

However, when applied to a system that is far from equilibrium GADFLI (GAlactic Dynamics From Lie-Poisson Integration) yields results that differ markedly from what is observed in N-body simulations [42]. Specifically, GADFLI does not yield the comparatively efficient approach towards equilibrium that is observed in simulations. For example, whereas N-body simulations of a collapsing spherical configuration typically exhibit oscillations which damp on a time scale $\sim t_{cr}$, the moment code yields oscillations that only damp appreciably on a significantly longer time scale. The obvious inference is that GADFLI is missing some important physical ingredient that is present in N-body codes. One possible interpretation [43] would be that standard N-body codes are contaminated by nonsymplectic perturbations that act as a spurious source of dissipation, and that a more satisfactory N-body code would exhibit a less efficient approach towards equilibrium. (N-body codes that use a time-centered leap frog scheme are in fact symplectic, but this scheme is not usually used for large N simulations which tend, generically, to exploit variable time steps.) However, a more conservative interpretation is that the N-body codes are at least approximately correct, and that GADFLI misses the effects of phase mixing which rely on the existence of a large number — formally an infinite number — of modes.

If one accepts tentatively the hypothesis that the flow associated with a self-gravitating system far from equilibrium should be interpreted in terms of nonlinear Landau damping, one is led to a single, unified picture of an evolution described by the CBE in which the single dominant mechanism is phase mixing/Landau damping. The idea is that, for a generic initial $f(0)$, the approach towards (albeit not to) an equilibrium will involve nonlinear oscillations about some time-independent f_0, oscillations which may — or may not — eventually phase mix away. When the system is far from equilibrium, nonlinearities will be important and mode-mode coupling will play an important role. Eventually, however, as the system approaches equilibrium more closely, nonlinearities will become less important and ordinary linear Landau damping/phase mixing will provide a reasonable description of what is actually happening.

Given this hypothesis, it is also clear what problem ideally one would like to solve: Determine how to break a generic $f(0)$ into two pieces, an equilibrium f_0 and a correction $\delta f(0)$ not necessarily small, such that, at least in the absence of any undamped modes, the perturbation $\delta\rho$ and/or $\delta\Phi$ associated with the evolved δf tends to zero at late times.

Another, slightly different, way in which to formulate this same picture, which may provide some insights into how to identify the equilibrium f_0 lurking in the initial $f(0)$, is the following: A Hamiltonian evolution governed by the CBE corresponds to a flow in an infinite-dimensional phase space which is constrained by conservation of phase (i.e., fixed values of all the Casimirs C) and the fact that the value of the mean field energy $\mathcal{H}$ does not change in time. Imposing these constraints explicitly serves to pick out a (still infinite-dimensional) hypersurface in the phase space to which the flow is restricted. By analogy with finite-dimensional Hamiltonian systems (cf. [44], [45]), one might expect that, when evolved into the future, an initial $f(0)$ will exhibit a coarse-grained approach towards some invariant measure on this hypersurface, for example, towards a suitably defined microcanonical population of the accessible phase space region. If, when viewed in the infinite-dimensional phase space, the flow is chaotic, so that nearby distribution functions f diverge exponentially, one might expect an approach towards an invariant measure which is exponential in time. If, alternatively, the flow is regular, one might expect a power law approach instead. However, in either case it would seem reasonable to expect an approach towards a "phase-mixed" microcanonical distribution proceeding on the natural time scale t_{cr}. In this context, the crucial questions would seem to be the following: for the fixed values of C associated with $f(0)$, what equilibrium solutions f_0 exist, and to what extent can the invariant distribution associated with the evolved $f(0)$ be interpreted as involving (nonlinear) oscillations about some lower energy f_0?

Weak Non-Hamiltonian Corrections to Collisionless Evolution

This entire picture relies crucially on the fact that the CBE is a Hamiltonian system. That the CBE is Hamiltonian is in turn a manifestation of a very general result, namely that, when suitably interpreted, the mean field description of *any* Hamiltonian system is itself Hamiltonian [46]. If, however, any corrections to mean field theory are introduced, one obtains generically a description of the one-particle distribution function f which is no longer Hamiltonian. Physically, this arises because one is treating the full N-particle distribution function μ as being a sum of two "pieces," namely (1) a product of identical one-particle distribution functions $f(i)$ ($i = 1, \ldots, N$) and (2) a correction reflecting particle-particle correlations which are ignored in the mean field description. The point then is that integrating over the correlations to obtain a closed equation for f leads to a nonlocal, non-Hamiltonian evolution equation [47].

At the level of particle trajectories, this intuition is manifested in the conventional description of binary encounters as resulting in diffusion and dynamical friction [2],

[6]. Thus, for example, for motion in a fixed potential, one can formulate evolution equations of the form

$$\frac{dx^a}{dt} = v^a \quad \text{and} \quad \frac{dv_a}{dt} = -\frac{\partial \Phi}{\partial x^a} - \eta(x^a, v_a)v_a + \mathcal{F}_a, \tag{12}$$

where η and $\mathcal{F}_a$ are related by a fluctuation-dissipation theorem, in terms of a characteristic "temperature" or mean squared velocity Θ. If, for example, one treats the fluctuating forces as Gaussian white noise with zero mean, η must be constant and (cf. [48])

$$\langle \mathcal{F}_a(t_1)\mathcal{F}_b(t_2)\rangle = 2\Theta\eta\delta_{ab}\delta_D(t_1 - t_2). \tag{13}$$

If the friction and noise are weak, they will only cause appreciable changes in the particle energy E (and any other quantities conserved in the Hamiltonian description) on a time scale $\gg t_{cr}$. However, this does not mean that they cannot have appreciable effects on the flow over much shorter time scales. Indeed, they can be important in at least two ways:

1. If the fixed potential is complicated, involving a coexistence of both regular and chaotic orbits, a strictly Hamiltonian evolution can result in chaotic orbits being trapped in certain phase space regions for a comparatively long time, even though, in the $t \to \infty$ limit, they will diffuse through cantori [49] or along an Arnold web [50] to probe a much larger phase space region [51], [52]. However, even very weak friction and noise — so weak that, over short time scales, they have negligible effects on the form of the invariant measure — tend generically to dramatically accelerate this diffusion, thus facilitating a much more rapid approach towards the invariant measure [52]–[54]. Thus, for example, in some cases friction and noise corresponding to a natural time scale as long as 10^6–$10^9 t_{cr}$ can have significant statistical effects on the evolution of ensembles of orbits on a time scale as short as $100 t_{cr}$. At least for orbits in a fixed potential, small non-Hamiltonian corrections can greatly facilitate the coarse-grained approach towards a statistical equilibrium.

2. Liouville's Theorem implies that trajectories in a Hamiltonian system cannot self-intersect. Thus, no matter how complicated the potential or the phase space, a Hamiltonian trajectory must evolve in such a fashion as to avoid crossing itself. However, this can lead to complicated microscopic structures, such as, for example, the "homoclinic tangle" that can arise near separatrices [55]. The important point then is that even very weak non-Hamiltonian perturbations can break the constraints associated with Liouville's Theorem, thus allowing the tangle to smooth itself out. In other words, small perturbations that break the rigid constraints associated with Liouville's Theorem can serve to fuzz out small scale structures that arise in a purely Hamiltonian description.

These two alternatives have been formulated completely in the context of motion in a fixed potential, and at present, little if anything is really known about the effects of small non-Hamiltonian perturbations in the context of a self-consistent description as provided by the CBE. However, by analogy with the simpler problem it would seem reasonable to conjecture that even very weak perturbations that break the constraints associated with conservation of phase could facilitate the approach towards a meta-equilibrium, and that they could help blur the complex structures which can arise microscopically as the system phase mixes towards this meta-equilibrium.

HOW DOES THE COLLISIONLESS BOLTZMANN EQUATION RELATE TO THE FULL *N*-BODY PROBLEM?

The gravitational N-body problem for a collection of particles of comparable mass is chaotic in the sense that a small initial perturbation will typically grow exponentially on a time scale $\sim t_{cr}$. This phenomenon, long known [56], has been studied systematically over the past several years [57]–[61], and the observed behavior has proven to be extremely robust. Specifically, one finds that the observed exponential growth is largely independent of the detailed choice of initial conditions and initial perturbation. Seemingly independent of the details, small initial perturbations grow exponentially until they become "macroscopic." Thus, in particular, this chaos persists even for a system that is in or near a statistical meta-equilibrium. Moreover, one finds that, when scaled in terms of an N-dependent crossing time t_{cr}, the time scale τ associated with this instability admits at most a weak dependence on N, at least for $N \gg 2$. In particular, the chaos does not appear to turn off for very large N. There is no sense in which $(\tau/t_{cr}) \to \infty$ for $N \to \infty$. If anything, the ratio τ/t_{cr} very slowly *decreases* with increasing N [60].

This leads, however, to an important question of principle. The N-body problem appears to be chaotic on a time scale $\sim t_{cr}$, but the flow associated with the CBE is often integrable or near-integrable in the sense that many or all of the characteristics are regular, i.e., nonchaotic. So what do the (often near-integrable) CBE characteristics have to do with the true (chaotic) N-body problem?

There is no *a priori* contradiction between a chaotic N-body problem and a CBE evolution that is integrable or near-integrable, since the CBE is predicated on the existence of a smooth distribution function f and a smooth potential Φ. However, there is at least one awkward point: Although galactic astronomers teach that CBE characteristics simply track the motion of phase space fluid elements, and that they should not be interpreted as representing the orbits of representative particles, there *is* oftentimes the implicit idea that, for $N \to \infty$, true N-body orbits will converge towards orbits in the smooth, self-consistent potential, i.e., towards CBE characteristics.

For a singular initial distribution function

$$f_N(x^a, v_a, 0) = \sum_{i=1}^{N} m\, \delta_D[\mathbf{x} - \mathbf{x}_i(0)]\, \delta_D[\mathbf{v} - \mathbf{v}_i(0)], \tag{14}$$

the CBE is of course equivalent to the full N-body problem. If, however, one considers a sequence of initial distribution functions with increasing number N, decreasing particle mass m, and constant $M = Nm$, there is no obvious sense in which, for N

$\rightarrow \infty$, the initial $f_N(0)$ converges towards a smooth distribution function $f_\infty(0)$. Rather, what one has is lower amplitude (but still infinite) spikes on progressively shorter scales. It is therefore hard to envision a uniform convergence towards a smooth potential or smooth CBE characteristics.

In this general context, it is also easy to understand qualitatively why chaos might be expected to persist even for very large N. Numerical simulations suggest that much, if not all, of the chaos observed in the N-body problem is associated with "random" interactions between nearest neighbors, perhaps amplified through a coupling to the bulk mean field. If, however, this is true, the time scale associated with the growth of an initial perturbation can be estimated by considering the tidal effects associated with the perturbation of a pair of particles separated by a distance comparable to a typical interparticle spacing. The tidal acceleration associated with a given pair will of course scale as

$$\delta\ddot{\mathbf{x}} = (\delta\mathbf{x}\cdot\nabla)\mathbf{a} \sim \frac{Gm}{r^3}\delta\mathbf{x}, \tag{15}$$

where r is their separation and m is a typical particle mass. Assuming, however, that $r \sim n^{-1/3}$, with n a typical number density, and that, at least roughly, $GNm/R_{sys} \sim \bar{v}^2$ with $\bar{v}$ a typical particle speed, it follows from dimensional analysis that the natural time scale $\tau \sim R_{sys}/\bar{v} \sim t_{cr}$. As N increases, the mass of the nearest neighbor decreases, but the distance to that neighbor also decreases in a fashion which cancels the effect of decreasing mass.

This leads to another awkward point. N-body simulators often "soften" their computations to reduce discreteness effects which, presumably, are more important in the small N systems which can actually be solved, than in the real, large N systems which one would like to understand. Thus, for example, it is customary in direct summation codes to replace the true $1/r$ kernel of Newtonian gravity by a new $1/\sqrt{r^2+\varepsilon^2}$ kernel which effectively turns off encounters on scales shorter than the softening length ε. However, if ε is chosen to be too large, i.e., not much smaller than the typical interparticle spacing, such a softening effectively kills the chaos which one expects to be present in the honest N-body problem even for very large N.

The correct answer to the question raised above — how to reconcile a chaotic N-body problem with a non-chaotic CBE flow — is not completely clear. What does, however, seem apparent from the preceeding is that, even for very large N, true N-body trajectories could differ significantly from CBE characteristics.

One interpretation of the CBE, which may enable one to finesse this question, is an "ensemble average" interpretation (cf. [62], [63]), in which any given realization of the N-body problem would be interpreted as involving the evolution of a collection of N initial conditions that randomly sample the smooth distribution function $f(0)$. In this context, there is no reason to expect that the orbits in an N-body realization will closely track characteristics of the CBE, but one might at least hope that, in some statistical sense, there is a close correspondence between N-body orbits and CBE characteristics.

To help focus on the basic issue, it is useful to consider an interesting thought experiment which, given recent advances in computer hardware, could perhaps actually be performed numerically. Start by specifying a smooth distribution function $f(0)$

and computing the characteristic associated with some phase point $(\mathbf{x}_0, \mathbf{v}_0)$. Next sample $f(0)$ to generate many different sets of initial conditions for the N-body problem, all with one particle at $(\mathbf{x}_0, \mathbf{v}_0)$, and evolve these different sets of initial conditions into the future. The obvious question is then: How do the N-body orbits generated from $(\mathbf{x}_0, \mathbf{v}_0)$ compare with the CBE characteristic, both individually and in terms of statistical averages?

Given the fact that the N-body problem is chaotic on a time scale $\sim t_{cr}$, it would seem reasonable to conjecture that the orbits generated in two different N-body realizations will diverge exponentially on a time scale $\sim t_{cr}$; and, similarly, one might expect that any given N-body orbit will diverge from the CBE characteristic on a comparable time scale. However, one might nevertheless expect that, for sufficiently large N, the ensemble average of the different N-body orbits generated from the same $(\mathbf{x}_0, \mathbf{v}_0)$ will closely track the CBE characteristic for some finite time. In particular, one might conjecture that the *rms* configuration space deviation between the N-body orbits and the CBE characteristic will scale as

$$\delta r_{rms}(t) \sim F(N) \exp(t/\tau), \tag{16}$$

where $\tau \sim t_{cr}$, roughly independent of the total particle number N, and where the prefactor $F(N) \rightarrow 0$ for $N \rightarrow \infty$. Thus, for example, the assumption of random fluctuations associated with a random choice of initial conditions might suggest $F(N) \propto 1/\sqrt{N}$.

Whether these expectations are correct is not clear. What does, however, seem evident is that galactic astronomers lack a clear understanding of the sense, if any, in which N-body orbits coincide approximately with CBE characteristics. This, however, is arguably an important lacuna. Much of the theorist's intuition derives from the study of orbits in smooth potentials presumably associated with CBE equilibria; and much of the standard interpretation of observational data is predicated on the assumption that stars follow simple nonchaotic orbits in a smooth gravitational potential.

ACKNOWLEDGMENTS

The ideas described here have derived in part from discussions — and arguments — with a number of different individuals, including George Contopoulos, Paul Channell, Salman Habib, Ted Kirkpatrick, Phil Morrison, and Daniel Pfenniger. I am grateful to Tom O'Neil, Dave Merritt, and Richard Lovelace for critical feedback on the talk from which this paper derives. I am especially grateful to Donald Lynden-Bell for his detailed criticisms of the penultimate version of the manuscript.

REFERENCES

1. MIHALAS, D. & J. BINNEY 1981. Galactic Astronomy. Freeman. San Francisco, CA.
2. BINNEY, J. & S. TREMAINE. 1987. Galactic Dynamics. Princeton University Press. Princeton, NJ.
3. SCHWEIZER, F. 1986. Colliding and merging galaxies. Science. **231:** 227–234.

4. MAOZ, E. 1991. What damps small perturbations in stellar systems. Astrophys. J. **375:** 687–690.
5. LYNDEN-BELL, D. 1967. Statistical mechanics of violent relaxation in stellar systems. Mon. Not. R. Astr. Soc. **136:** 101–121.
6. CHANDRASEKHAR, S. 1943. Principles of Stellar Structure. University of Chicago Press. Chicago, IL.
7. KANDRUP, H.E. 1981. A generalized Landau equation for a system with a self-consistent mean field: Derivation from an N-particle Liouville equation. Astrophys. J. **244:** 316–330.
8. HERTEL, P. & W. THIRRING. 1971. Free energy of gravitating fermions. Commun. Math. Phys. **24:** 22–36.
9. RYBICKI, G. 1971. Exact stastistical mechanics of a one-dimensional self-gravitating system. Astrophys. Space Sci. **14:** 56–72.
10. MINEAU, P., M.R. FEIX & J.L. ROUET 1990. Numerical simulations of violent relaxation and formation of phase space holes in gravitational systems. Astron. Astrophys. **228:** 344–349.
11. LUWEL, M. & G. SEVERNE. 1985. Collisions mixing in 1-dimensioanl gravitational systems initially in a stationary waterbag configuration. Astron. Astrophys. **152:** 305–314.
12. YAWN, K.R. & B.N. MILLER. 1997. Ergodic properties and equilibrium of one-dimensional self-gravitating systems. Phys. Rev. E. **56:** 2429–2436.
13. MORRISON, P.J. 1980. the Maxwell-Vlasov equations as a continous Hamiltonian system. Phys. Lett. A. **80:** 383–386.
14. KANDRUP, H.E. 1990. Geometric approach to secular and nonlinear stability for spherical star clusters. Astrophys. J. **351:** 104–113.
15. MORRISON, P.J. & S. ELIEZER. 1986. Spontaeous symmetry breaking and neutral stability in the noncanonical Hamiltonian formalis. Phys. Rev. A. **33:** 4205–4214.
16. BARTHOLOMEW, P. 1971. On the theory of stability of galaxies. Mon. Not. R. Astr. Soc. **151:** 333–350.
17. PFAFFELMOSER, K. 1992. Global classical solutions of the Vlasov-Poisson system in three dimensions for general initial data. J. Diff. Eqns. **95:** 281–303.
18. SCHAEFFER, J. 1991. Global existence of smooth solutions to the Vlasov-Poisson system in three dimensions. Commun. Part. Diff. Eqns. **16:** 1313–1335.
19. KORMENDY, J. & D. RICHSTONE. 1995. Inward bound — the search for supermassive black holes in galactic nuclei. Ann. Rev. Astron. Astrophys. **33:** 581–624.
20. BATT, J. 1987. Asymptotic properties of spherically symmetric self-gravitating systems for $t \rightarrow \infty$. Transport Theory and Statistical Physics **16:** 763–778.
21. LOUIS, P. D. & O.E. GERHART. 1988. Can galaxies oscillate? A self-consistent model of a non-stationary stellar system. Mon. Not. R. Astr. Soc. **233:** 337–365.
22. SRIDHAR, S. 1989. Undamped oscillations of homogeneous collisionless stellar systems. Mon. Not. R. Astr. Soc. **201:** 1159–1163.
23. KING, I. 1962. The structure of star clusters. I. An empirical density law. Astron. J. **67:** 471–485.
24. DE VAUCOULEUR, G. 1948. Recherches sur les nébuleuses extragalactiques. Ann. D'astrophysique. **11:** 247–287.
25. LAUER, T.R., E.A. AJHAR, Y.-L. BYUN, A. DRESSLER, S.M. FABER, C. GRILL-MAIR, J. KORMENDY, D. RICHSTONE & S. TREMAINE. 1995. The centers of early-type galaxies with *HST*. I. An observational study. Astron. J. **110:** 2622–2654.
26. LANDAU, L.D. & E.M. LIFSHITZ.1958. Statistical Physics. Pergamon. London.
27. IPSER, J.R. & G. HORWITZ. 1979. The problems of maximizing functionals in Newtonian stellar dynamics and its relation to thermodynamic and dynamical stability. Astrophys. J. **232:** 863–873.

28. HOLM, D.D., G.E. MARSDEN, T. RATIU & A. WEINSTEIN. 1985. Nonlinear stability of fluid and plasma equilibria. Phys. Repts. **123:** 1–116.
29. KANDRUP, H.E., M.E. MAHON & H. SMITH. 1993. Energy and phase space mixing for self-gravitating systems of stars. Astron. Astrophys. **271:** 440–450.
30. CONTOPOULOS, G. 1994. Private communication.
31. VAN ALBADA, T.S. 1982. Dissipationless galaxy formation and the $r^{1/4}$ law. Mon. Not. R. Astr. Soc. **201:** 939–955.
32. QUINN, P. J. & W.H. ZUREK. 1988. The angular momentum distribution in galactic halos. Astrophys. J. **331:** 1–18.
33. KANDRUP, H.E., M.E. MAHON & H. SMITH. 1993. Energy and phase space mixing for self-gravitating systems of stars. Astron. Astrophys. **271:** 440–450.
34. KANDRUP, H.E. 1993. Non-violent relaxation of colliding galaxies. *In* Advances In Gravitation and Cosmology. B.R. Iyer, A.R. Prasanna, R.K. Varma & C.V. Vishveshwara, Eds. Wiley Eastern. New Delhi.
35. LANDAU, L.D. 1946. On the vibrations of the electronic plasma. Soviet Phys. JETP. **10:** 25.
36. STIX, T.H. 1962. The Theory Of Plasma Waves. Mcgraw-Hill. New York.
37. CASE, K.M. 1959. Plasma oscillations. Ann. Phys. (NY) **7:** 349–364.
38. VAN KAMPEN, N.G. 1955. On the theory of stationary waves in plasmas. Physica **21:** 949–963.
39. HABIB, S., H.E. KANDRUP & P.F. YIP. 1986. Gravitational Landau damping. Astrophys. J. **309:** 176–182.
40. MONTGOMERY, D. 1963. Nonlinear Landau damping of oscillations in a bounded plasma. Phys. Fluids A . **6:** 1109–1114.
41. CHANNELL, P.J. 1995. Canonical integration of the collisionless Boltzmann equation. Ann. N. Y. Acad. Sci. **751:** 152–161.
42. KANDRUP, H.E. 1994. Is There Chaos In BEDLAM? University of Florida preprint.
43. CHANNELL, P.J. 1994. Private communication.
44. KANDRUP, H.E. & M.E. MAHON. 1994. Relaxation and stochasticity in a truncated Toda lattice. Phys. Rev. E . **49:** 3735–3747.
45. MAHON, M.E., R.A. ABERNATHY, B.O. BRADLEY & H.E. KANDRUP. 1995. Transient ensemble dynamics in time-independent galactic potentials. Mon. Not. R. Astr. Soc. **275:** 443–453.
46. KANDRUP, H.E. 1994. Reduced Hamiltonian descriptions. Phys. Rev. D. **50:** 2425–2430.
47. KUBO, R., M. TODA & N. HASHITSUME. 1991. Statistical Physics II. Springer-Verlag. New York.
48. CHANDRASEKHAR, S. 1943. Stochastic problems in physics and astronomy. Rev. Mod. Phys. **15:** 1–89.
49. MATHER, J.N. 1982. Existence of quasi-periodic orbits for twist homeomorphisms of the annulus. Topology. **21:** 457–467.
50. ARNOLD, V.I. 1964. Small denominators and problems of stability of motion in classical and celestial mechanics. Russian Math. Surveys. **18:** 85–191.
51. CONTOPOULOS, G. 1971. Orbits in highly perturbed dynamical systems. Astron. J. **76:** 147–156.
52. LIEBERMAN, M.A. & A. LICHTENBERG. J. 1972. Stochastic and adiabatic behaviour of particles accelerated by periodic forces. Phys. Rev. A **5:** 1852–1866.
53. HABIB, S., H.E. KANDRUP & M.E. MAHON. 1996. Chaos and noise in a truncated Toda potential. Phys. Rev. E. **53:** 5473–5476.
54. HABIB, S., H.E. KANDRUP & M.E. MAHON. 1997. Chaos and noise in galactic potentials. Astrophys. J. **480:** 155–166.

55. Lichtenberg, A.J. & M.A. Lieberman. 1992. Regular and Chaotic Dynamics. Springer-Verlag. New York.
56. Miller, R.H. 1964. Irreversibility in small stellar dynamical systems. Astrophys. J. **140:** 250–256.
57. Kandrup, H.E. & H. Smith. 1991. On the sensitivity of the *N*-body problem to small changes in initial conditions. Astrophys. J. **374:** 255–265.
58. Kandrup, H.E. & H. Smith. 1992. On the sensitivity of the *N*-body problem to small changes in initial conditions. II. Astrophys. J. **386:** 635–645.
59. Kandrup, H.E., H. Smith. & D.E. Willmes. 1992. On the sensitivity of the *N*-body problem to small changes in initial conditions. III. Astrophys. J. **399:** 627–633.
60. Goodman, J., D. Heggie & P. Hut. 1994. On the exponential instability of *N*-body systems. Astrophys. J. **415:** 715–723.
61. Kandrup, H.E., M.E. Mahon & H. Smith. 1992. On the sensitivity of the *N*-body problem to small changes in initial conditions. III. Astrophys. J. **428:** 458–465.
62. Ogorodnikov, K.F. 1965. Dynamics of Stellar Systems. Macmillan. New York.
63. Klimontovich, Yu. L. 1983. The Kinetic Theory of Electromagnetic Processes. Springer-Verlag. New York.
64. Lynden-bell, D. & R. Wood. 1968. The gravo-thermal catastrophe in isothermal spheres and the onset of Red-Giant structure for stellar systems. Mon. Not. R. Astr. Soc. **138:** 495–525.

Self-Consistent Gravitational Chaos[a]

DAVID MERRITT AND MONICA VALLURI

Department of Physics and Astronomy, Rutgers University, New Brunswick, New Jersey 08855

ABSTRACT: The motion of stars in the gravitational potential of a triaxial galaxy is generically chaotic. However, the timescale over which the chaos manifests itself in the orbital motion is a strong function of the degree of central concentration of the galaxy. Here, chaotic diffusion rates are presented for orbits in triaxial models with a range of central density slopes and nuclear black-hole masses. Typical diffusion times are found to be less than a galaxy lifetime in triaxial models where the density increases more rapidly than $\sim r^{-1}$ at the center, or which contain black holes with masses that exceed ~0.1% of the galaxy mass. When the mass of a central black hole exceeds roughly 0.02 $M_{\rm gal}$, there is a transition to global stochasticity and the galaxy evolves to an axisymmetric shape in little more than a crossing time. This rapid evolution may provide a negative feedback mechanism that limits the mass of nuclear black holes to a few percent of the stellar mass of a galaxy.

1. INTRODUCTION

The gravitational force that determines the motion of a star in a galaxy has two components: the smooth force generated by the overall mass distribution, and the nonsmooth force that results from close encounters between individual stars. The relative importance of the two components after one orbital period is roughly equal to $0.1N/\log N$, where N is the number of stars in the galaxy [1]. For a typical galaxy with $N \approx 10^{11}$ stars, close encounters between stars are unimportant and the dynamics are essentially collisionless, at least over timescales of 10^2–10^3 orbital periods that correspond to galaxy lifetimes.

Motion in a smooth gravitational field becomes quite simple if the number of isolating integrals equals or exceeds the number of degrees of freedom. Much work in galactic dynamics has focussed on finding integrable or near-integrable models for galactic potentials [2]. Kuzmin [3], [4] showed that there is a unique, ellipsoidally-stratified mass model for which the corresponding potential has three global integrals of the motion, quadratic in the velocities. Kuzmin's model — explored in detail by de Zeeuw [5], who christened it the "Perfect Ellipsoid" — has a large, constant-density core in which the orbital motion is that of a 3-D harmonic oscillator. Every orbit in the core of the Perfect Ellipsoid fills a rectangular parallelepiped, or "box." At higher energies in the Perfect Ellipsoid, box orbits persist, and three new orbit families appear: the "tubes," orbits that preserve the direction of their motion around either the long or short axis of the figure. Tube orbits respect an integral of the motion analogous to the angular momentum, and hence — unlike box orbits — avoid

[a] This work was supported by NASA grant NAG 5-2803 and by NSF grants AST 93-18617 and AST 96-17088.

the center. When two of the axis lengths of an ellipsoidal model are equal, the box orbits disappear, and all trajectories belong to a single family of tube orbits that circulate about the axis of symmetry. Thus, box orbits are uniquely associated with the triaxial geometry.

Unfortunately, the Perfect Ellipsoid does not look very much like real elliptical galaxies. Its density falls off as $\rho \propto r^{-4}$ at large radii; both the luminosity and mass density in real galaxies fall off more slowly, as r^{-3} or r^{-2}. And recent observations of galactic nuclei demonstrate that constant-density cores do not exist — the density of starlight in real galaxies always rises monotonically at small radii, roughly as a power law [6]. The density profile in these power-law "cusps" rises as steeply as $\rho \propto r^{-2}$ in faint elliptical galaxies, while the cusps in brighter galaxies are typically shallower [7]. There is also increasingly strong evidence that many elliptical galaxies and bulges contain central massive objects, possibly the black holes that are thought to have powered quasars [8]. While the masses of these dark central components are often very uncertain, typical estimates are $10^{-3} \lesssim M_{BH}/M_{\rm gal} \lesssim 10^{-2}$, where M_{BH} is the black hole mass inferred from the orbital motions of surrounding stars and gas, and $M_{\rm gal}$ is the stellar mass of the host galaxy or (in the case of a spiral galaxy) the mass of the stellar bulge. Some galaxies, like M32, the dwarf companion to the nearby Andromeda galaxy, are known to contain both a steep stellar cusp ($\rho \propto r^{-1.6}$) and a dynamically-significant black hole ($M_{BH}/M_{\rm gal} \sim 0.003$).

The existence of box orbits in the Perfect Ellipsoid is tied to the stability of the long- or x-axis orbit [9]. The x-axis orbit is unstable at most energies to lateral perturbations in triaxial models where the density increases more rapidly than $\sim r^{-1}$ near the center [10]. The instability first appears through the bifurcation of a 1:2 resonant orbit, the $x - z$ "banana" boxlet [11]. The x-axis orbit is likewise unstable in any triaxial model with a central singularity [12]. It follows that bona-fide box orbits do not exist in the majority of triaxial potentials corresponding to real elliptical galaxies; in their place, we would expect to find either stochastic orbits, or regular orbits associated with minor resonances (like the banana) that avoid the center.

The nonexistence of box orbits has important consequences for the self-consistent dynamics of elliptical galaxies. Schwarzschild [13], [14] and Statler [15] found that box orbits — particularly the thin boxes that remain close to the long axis — were crucial for reconstructing the distribution of mass in triaxial models. Their work was based on mass models with smooth cores. Merritt and Fridman [16] attempted to construct selfconsistent triaxial models with central density cusps, after excluding the stochastic boxlike orbits, or replacing them with invariant ensembles representing a uniform population of stochastic phase space. Completely stationary solutions could not be found; only quasi-equilibrium solutions, in which stochastic phase space was populated in a nonuniform way, could successfully reproduce the density at all points in the model. An extension of this work to triaxial models with a range of axis ratios [17] revealed that stationary models with r^{-2} cusps could only be constructed if the figure was nearly axisymmetric.

One important question left unanswered by these equilibrium studies is the timescale over which chaos implies changes in the self-consistent structure of a galaxy. Here, results from two recent studies that address this question [18], [19] are presented.

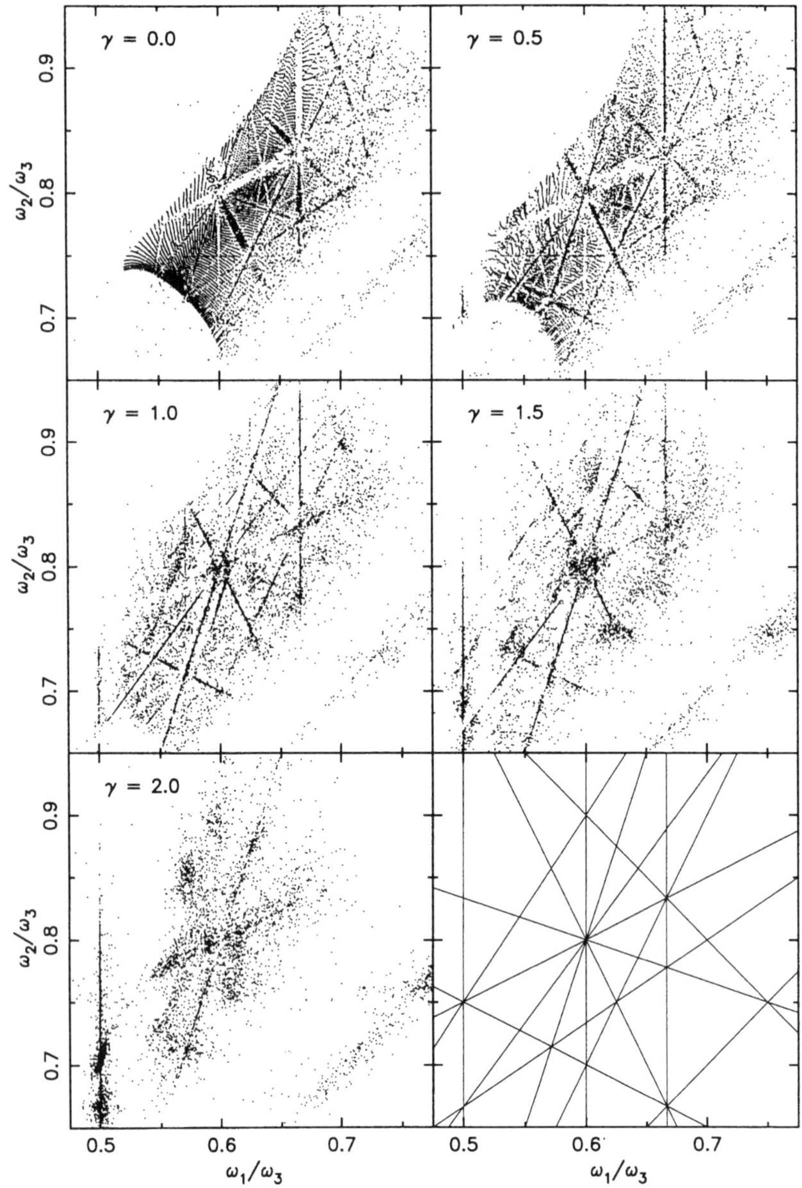

FIGURE 1. Frequency maps of boxlike orbits in triaxial potentials corresponding to the mass distribution of Eq. (2), with various cusp slopes γ [18]. ω_1, ω_2, and ω_3 are the fundamental frequencies associated with oscillations along the long, intermediate, and short axes of the triaxial ellipsoid. The nearly regular grid of points in the upper left-hand panel indicates that most of the orbits in the $\gamma = 0$ model mimic regular orbits over 50 oscillations. Departures from a regular grid imply that diffusion has taken place in frequency space, i.e., that the motion is stochastic. The bottom right panel shows the most important resonance layers, defined by $l\omega_1 + m\omega_2 + n\omega_3 = 0$ with integer $\{l, m, n\}$.

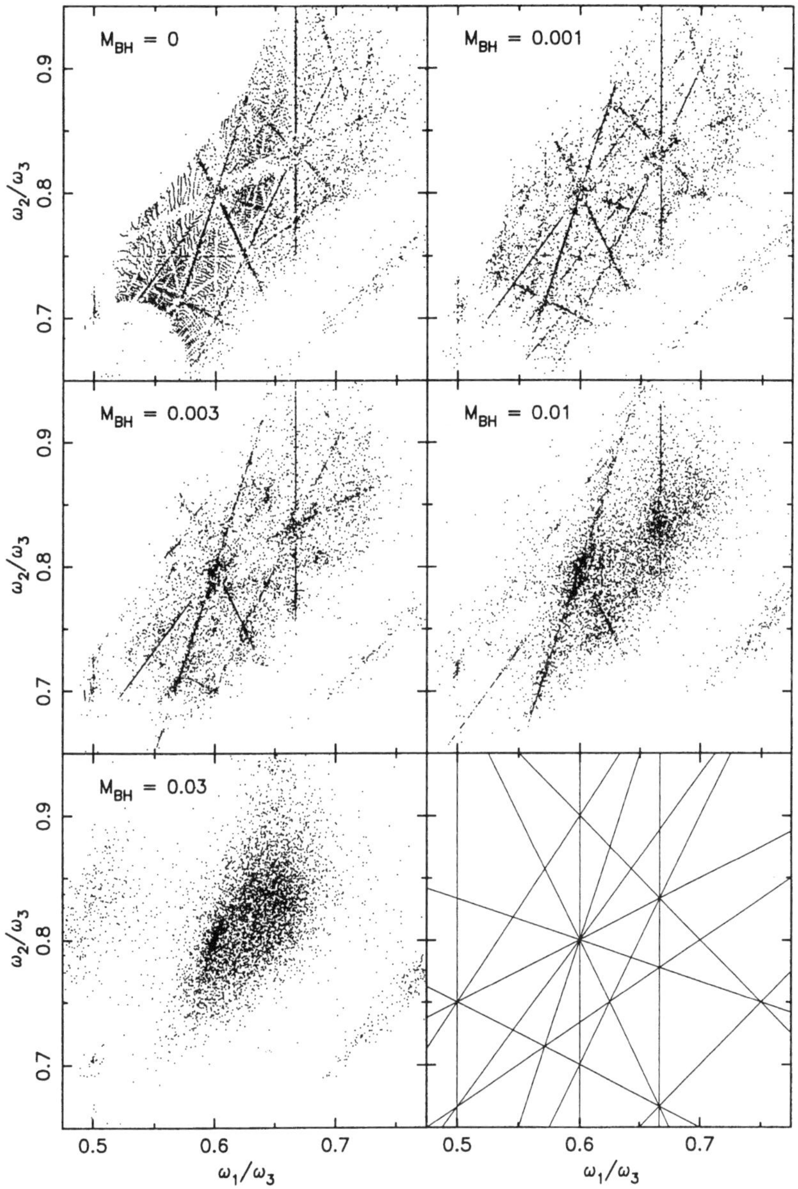

FIGURE 2. Like FIG. 1, for boxlike orbits in Dehnen's model (Eq. (2)) with $\gamma = 0.5$ and with an added central point mass containing various fractions M_{BH} of the total mass of the model [18].

DIFFUSION OF STOCHASTIC ORBITS

In order for chaos to be relevant to real galaxies, it must produce a significant change in the region visited by an orbit after just a few tens or hundreds of orbital periods — the approximate lifetime of a galaxy. In a pioneering study, Goodman and Schwarzschild [9] found that the boxlike orbits in a triaxial model with a smooth core were often stochastic, but that the orbital motion was essentially regular for at least 50 oscillations. On the other hand, Merritt and Fridman [16] found that the stochasticity in triaxial models with $\rho \propto r^{-2}$ density cusps produced significant changes in the appearance of boxlike orbits after just a few tens of oscillations. Merritt and Valluri [20] went on to calculate timescales for mixing in these strongly chaotic potentials; they found that ensembles of stochastic trajectories evolved toward invariant distributions — corresponding to an approximately uniform filling of stochastic phase space — on timescales of only 10^1–10^2 orbital periods. Taken together, these results suggest that the characteristic time over which chaos manifests itself in the orbital motion is strongly dependent on the central concentration of a triaxial model. Presumably, this dependence reflects the sensitivity of boxlike orbits to deflections that occur during close passages to the galaxy center.

A useful way to quantify the diffusion of stochastic orbits has been described by Laskar [21], [22]. His "frequency mapping" technique is based on the fact that a regular orbit can be characterized by the three fundamental frequencies ω_i, $i = 1, 2, 3$ that define motion on an invariant torus. These frequencies can be computed with high accuracy by integrating an orbit for a finite length of time ΔT, storing its phase-space coordinates at regular intervals, and decomposing the motion into Fourier components. Carrying out this procedure for a large number of orbits at a given energy and plotting ω_1/ω_3 vs. ω_2/ω_3 gives the "frequency map," a regular grid of points in frequency space. The same procedure can be carried out for stochastic orbits, but now the values of the ω_i will depend on the integration interval — a stochastic orbit is not confined to a single invariant torus and its "characteristic frequencies" will change with time as it diffuses along the energy surface. A strongly chaotic orbit will migrate through frequency space in just a few orbital periods, while a weakly chaotic orbit can remain close to a single invariant torus for a very long time. The diffusion is apparent as a distortion of the frequency map; the size of the distortion is proportional to the distance that the orbits have moved in frequency space during the period of integration ΔT.

Papaphilippou and Laskar [23], [24] demonstrated the usefulness of the frequency mapping technique for galaxy dynamics by using it to study orbital motion in the logarithmic potential,

$$\Phi(x, y, z) = \log(m^2 + r_c^2), \qquad m^2 = \frac{x^2}{a^2} + \frac{y^2}{b^2} + \frac{z^2}{c^2}. \tag{1}$$

Equation (1) is the gravitational potential generated by a triaxial mass model in which the density falls off as r^{-2} outside of the core radius r_c. Papaphilippou and Laskar established that the motion of boxlike orbits in the logarithmic potential is generically chaotic, and showed that much of the chaos is associated with motion out of the principal planes.

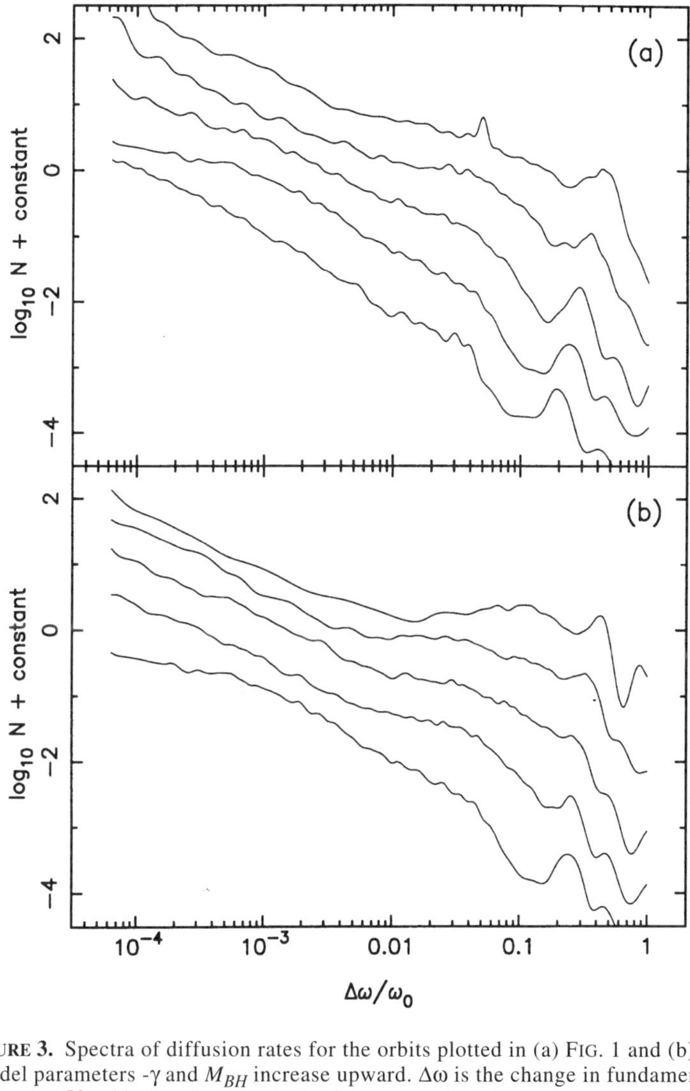

FIGURE 3. Spectra of diffusion rates for the orbits plotted in (a) FIG. 1 and (b) FIG. 2. The model parameters -γ and M_{BH} increase upward. $\Delta\omega$ is the change in fundamental frequencies over 50 orbital periods; ω_0 is the frequency of the long-axis orbit.

A mass model that better represents the distribution of starlight near the centers of elliptical galaxies and bulges is Dehnen's law [25],

$$\rho(m) = \rho_0 m^{-\gamma}(1 + m)^{-(4-\gamma)}. \tag{2}$$

Dehnen's formula allows the steepness of the central density cusp to be adjusted through the parameter γ — a useful feature, since real elliptical galaxies and bulges have cusps with a range of slopes, $0 \lesssim \gamma \lesssim 2$. FIGURE 1 shows frequency maps of boxlike orbits in triaxial potentials derived from Dehnen's density law, with $c/a = 1/2$ and $b/a = 0.79$, and five different values of γ. All the orbits in a given map have the same energy, corresponding roughly to the gravitational potential at the half-mass radius. The fundamental frequencies were computed over a time interval ΔT equal to 50 periods of the x-axis orbit — roughly a galaxy lifetime. Approximately 10^4 orbits were integrated for each map; the initial conditions consisted of a grid of points distributed uniformly over the equipotential surface [18].

The frequency maps of FIG. 1 show a clear transition from nearly regular behavior over 50 orbital periods when $\gamma = 0$ or 0.5, to clearly chaotic motion when $\gamma = 1.5$ or 2. Many of the orbits can be seen to lie in resonance zones, regions associated with rational resonant layers $l\omega_1 + m\omega_2 + n\omega_3 = 0$, which appear as lines in the frequency maps. The last resonance to strongly influence the motion as γ approaches 2 is the $1:2$ $x-z$ banana resonance, which generates the vertical lines $\omega_1/\omega_3 = 0.5$ in FIG. 1. For smaller values of γ the motion is influenced by a number of different resonances, some of rather high order.

By computing the fundamental frequencies over two adjacent time intervals, each of length ΔT, one can define a rate of diffusion in frequency space $\Delta\omega/\Delta T = \max\{|\Delta\omega_i|, i = 1, 2, 3\}/\Delta T$ [26]. The distribution of $\Delta\omega$'s is shown in FIG. 3 for each set of orbits used in the construction of the frequency maps of FIG. 1. Remarkably, there are no separate peaks associated with regular orbits, i.e., orbits with constant ω's (such orbits would be expected to have$\Delta\omega/\omega_0 \approx 10^{-4}$, the approximate accuracy of the numerical routine that calculates the fundamental frequencies). Instead, the spectrum of diffusion rates continues to rise toward small $\Delta\omega$, roughly as a power law. While there undoubtedly exist regular boxlike orbits associated with stable resonances like the banana, FIG. 3 suggests that the phase space volume associated with such orbits is very small. As γ approaches zero, the slope of the distribution increases, i.e., a smaller fraction of the boxlike orbits exhibit strong diffusion. But even for $\gamma = 0$ there is no hint of a separate population of regular orbits — rather, the typical diffusion times become much longer than 50 orbital periods, indicating that the majority of boxlike orbits are trapped for long periods of time in narrow regions of phase space.

These results suggest that the motion of boxlike orbits in triaxial potentials is generically chaotic, but that the distribution of diffusion rates varies systematically with the degree of central concentration of the model. In models with low central concentrations, typical diffusion times are long compared to 50 orbital periods, but as the central concentration is increased, a larger fraction of the orbits are able to diffuse significantly in a galaxy lifetime. A very crude index of the importance of the diffusion can be defined as the fraction of boxlike orbits for which $\Delta\omega/\omega_0 > 0.1$, i.e., the fraction of orbits which experience a 10% or greater change in their fundamental frequencies over 50 orbital periods. FIGURE 4 shows that this fraction increases from ~10% at $\gamma = 0$ to ~20% at $\gamma = 1$ and ~50% at $\gamma = 2$. Thus, a typical boxlike orbit is essentially regular over the lifetime of a galaxy in triaxial potentials with $\gamma \lesssim 1$, and essentially chaotic when $\gamma \approx 2$. This conclusion is consistent with the very different configuration-space appearance of boxlike orbits integrated in triaxial potentials with cores [9] and with steep cusps [16].

The effect of a central singularity on the frequency maps is shown in FIG. 2, computed from Dehnen's $\gamma = 0.5$ model with an added central point mass. When the

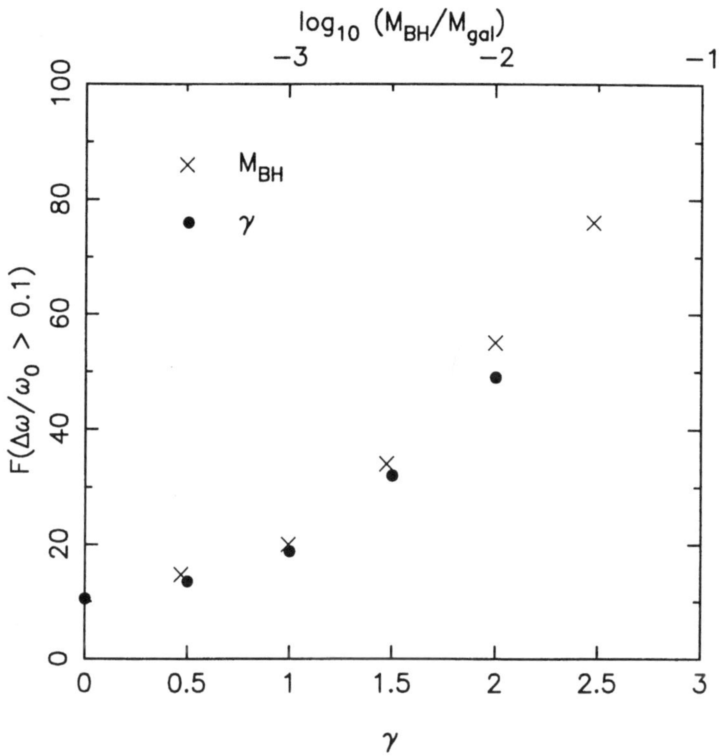

FIGURE 4. Fraction of boxlike orbits that undergo strong diffusion, $\Delta\omega/\omega_0 \geq 0.1$, in 50 orbital periods.

black hole mass M_{BH} is less than about 0.1% of the galaxy mass, the frequency map differs only slightly from that of the model with $M_{BH} = 0$. As M_{BH} is increased to 1%, the motion becomes clearly chaotic, and for $M_{BH} = 3\%$ there is hardly a trace of structure remaining in the frequency map. FIGURE 3 verifies that the spectrum of diffusion rates becomes very shallow for large M_{BH}, even increasing toward large $\Delta\omega$ when $M_{BH} \gtrsim 0.3\%$. The fraction of boxlike orbits undergoing strong diffusion is illustrated in FIG. 4; for $M_{BH} = 3\%$, fully 3/4 of the orbits evolve strongly over 50 orbital periods.

The frequency maps paint a more complex picture of the phase space of triaxial potentials than was apparent from earlier studies based on cruder techniques for characterizing the stochasticity. For instance, instability timescales of boxlike orbits inferred from Liapunov exponents are only 3–5 orbital periods in triaxial models like those studied here, even in models with smooth cores [20]. This is much shorter than the typical diffusion time inferred from the frequency map analysis. The apparent discrepancy can be understood by recognizing that the motion in weakly chaotic potentials is confined over long periods of time to narrow regions in phase space. The Liapunov exponents measure only the divergence rate of nearby trajectories within these limited regions, and not the physically more interesting timescale for diffusion from one such region to another. As the degree of central concentration of the model is increased, the stochastic layers increase in size until they overlap, and a large fraction of the boxlike orbits are able to wander ergodically over the energy surface in a finite length of time. This change in the structure of the phase space is reflected in the spectrum of diffusion rates, which becomes shallower as γ or M_{BH} are increased.

Dynamical systems often exhibit a transition to "global stochasticity:" as a perturbation parameter is increased, there is a sudden change from a regime in which the stochastic motion is closely bounded by KAM surfaces, to a regime where the stochastic motion is interconnected over large portions of the phase space. (Stochastic phase space in a 3 DOF system is always interconnected through the Arnold web, but Arnold diffusion is extremely slow unless the stochastic regions overlap.) In the globally-stochastic regime, there are few barriers to the motion, and stochastic orbits can wander over the full energy surface in little more than an orbital period. Is there a transition to global stochasticity in the triaxial models discussed here? Hints of such a transition can be seen in FIG. 4, which shows that the fraction of boxlike orbits that evolve strongly in 50 orbital periods becomes large as M_{BH} is increased beyond about $0.01M_{\text{gal}}$. In fact the situation is even more dramatic than this, as illustrated in FIG. 5, which shows the behavior of boxlike orbits integrated for just 10 orbital periods in triaxial models with a smooth core and various values of M_{BH}/M_{gal}. When the black hole mass exceeds ~2% of the galaxy mass, orbital evolution takes place in only a few oscillations — just the behavior expected in the globally- stochastic regime. If one imagines slowly increasing the mass of a black hole at the center of a triaxial galaxy, FIG. 5 suggests that the galaxy would be forced to respond rapidly — perhaps in just a few crossing times — once the black hole mass exceeded ~0.02 M_{gal}.

SELF-CONSISTENT EVOLUTION

In recent studies of the triaxial self-consistency problem [16], [17], [27], attempts were made to identify the stochastic orbits and to treat them differently from the regular orbits in the model construction. The separation of regular from stochastic orbits in these studies was based on the detection of linear instability of the motion; regular orbits were defined simply as those that showed no clear evidence of instability over ~100 oscillations. But the true situation is more complex, as discussed above. Boxlike orbits exhibit a wide range of diffusion rates, with no clear separation into "regular" and "stochastic" families. In addition, orbital periods decrease toward the center of a galaxy; an orbit with a diffusion time of 10^3 periods would mimic a regular orbit in the outer parts of a galaxy, but would behave chaotically over a galaxy lifetime if located near the center.

There would seem to be no substitute for N-body codes when dealing with a situation as complex as this. Until recently, N-body algorithms were unable to deal in a practical way with the high degree of central concentration of realistic triaxial models. But the situation has changed, and a number of algorithms are now available that can efficiently represent the gravitational potentials of systems with central

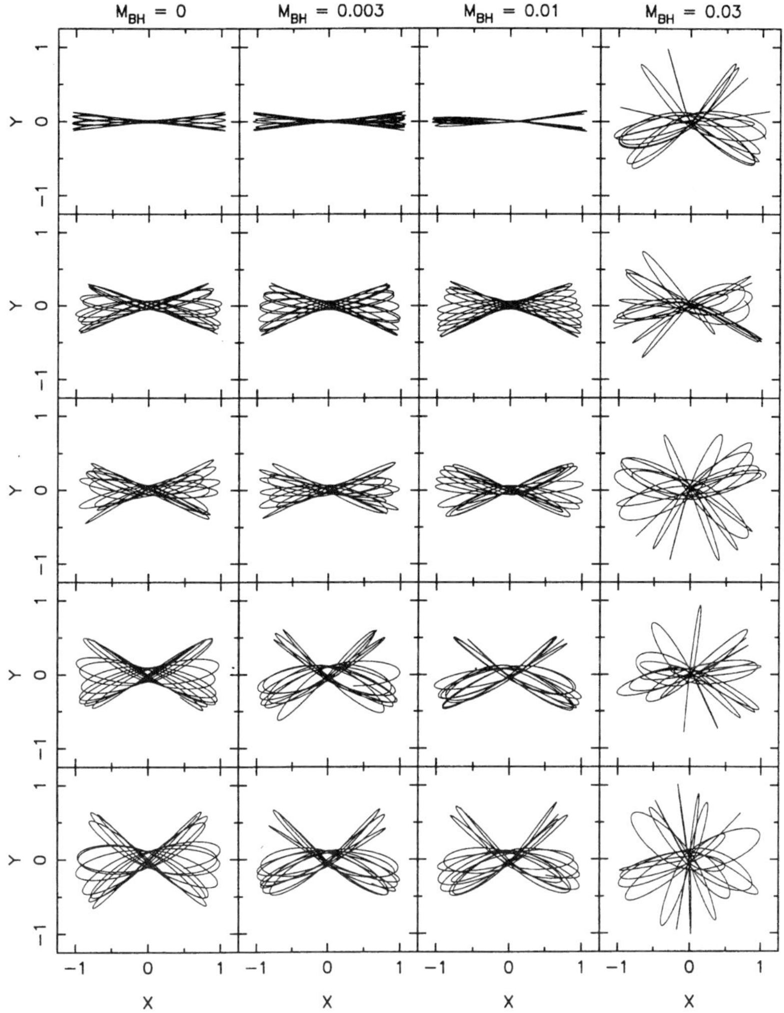

FIGURE 5. Test-particle integrations in a triaxial model with a smooth core and a central black hole. The z-axis is the short axis of the figure. Each of the five orbits is defined by its starting point on the equipotential surface, which remained fixed in angular position as M_{BH} was increased. Orbits were integrated for approximately 10 full oscillations. Chaos induces substantial changes in the orbital shapes over this period of time when $M_{BH} \gtrsim 0.02 M_{\rm gal}$.

cusps and nuclear black holes [28]. Here, the results of a new N-body study [19] of triaxial galaxies with central singularities is presented.

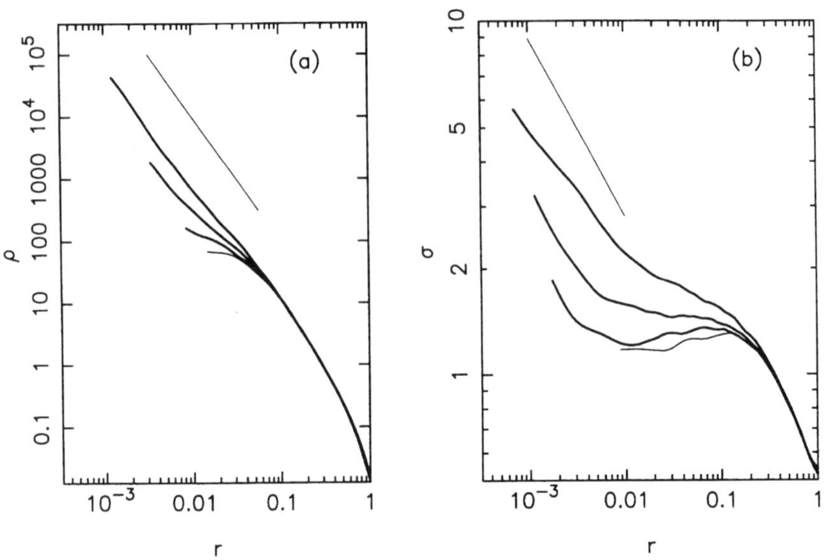

FIGURE 6. Growth of central density cusps in N-body models with various black hole masses [19]. (a) Density profiles; (b) velocity dispersion profiles. The thin curve is the initial model without a black hole; the three heavy curves are final models with $M_{BH}/M_{\text{gal}} = 0.003, 0.01, 0.03$. The straight line in panel (a) has a logarithmic slope of ~2; in panel (b), of ~0.5.

The initial conditions for this study consisted of $\sim 10^5$ particles distributed in a nonrotating triaxial bar, with $c/a \approx 0.5$ and $b/a \approx 0.75$. The initial model was generated by collapsing a cold, spherical cloud and allowing it to evolve to an equilibrium state; the triaxial shape resulted from a bar-making instability associated with cold collapse [29]. In order to increase the resolution very near the center, particles were given masses that depended on their initial positions; masses varied by a factor of 10 from the center to the envelope. Particle positions were advanced with individual timesteps, sometimes differing by as much as a factor of 10^6, using a fourth-order integrator.

The initial model had a constant-density core. A "black hole" was grown in this core by increasing the mass at the origin according to $M(t) = M_{BH}\tau^2(3 - 2\tau)$, with $\tau = t/t_{\text{grow}}$. Various values for M_{BH} and t_{grow} were used. FIGURE 6 illustrates the formation of a density cusp in the stars surrounding the black hole; the cusp forms because the black hole pulls in the neighboring stars as the force of its attraction increases [30].

The growth of a black hole in an initially triaxial model is expected to make the model more axisymmetric, through a two-step process [12], [31]. Many of the box orbits in the initial model are rendered stochastic by the black hole; as a result, they evolve to fill a more-or-less spherical volume corresponding roughly to the region enclosed by an equipotential surface. Nearly-spherical orbits are not very useful for

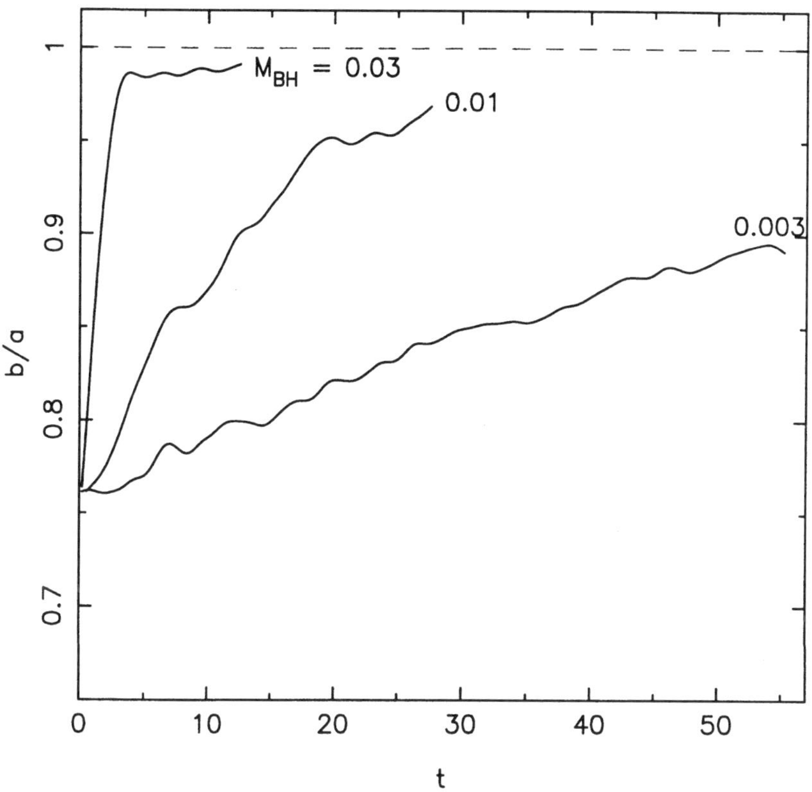

FIGURE 7. Evolution of the intermediate-to-long axis ratio *b/a* of the *N*-body models as nuclear black holes of three different masses are grown [19]. Time is in units of the half-mass orbital period. Black hole masses are expressed in terms of the total mass of the galaxy model. The axis ratio was computed from the most-bound 50% of the stars. The two smaller black holes were grown on a timescale of 5 half-mass orbital periods; the black hole with $M_{BH} = 0.03$ was grown in 2 half-mass orbital periods.

reconstructing a barlike shape, and so the model responds by becoming more axisymmetric. As it approaches axisymmetry, the tube orbits that circulate about the symmetry axis are able to reproduce the mass distribution self-consistently and the galaxy settles rapidly into equilibrium.

This picture appears to be essentially correct, as illustrated in FIG. 7, which shows the evolution of the intermediate-to-short axis ratio *b/a* of the model as the black hole is grown. For three different black hole masses, $M_{BH}/M_{gal} = 0.3\%$, 1%, and 3%, the model evolves to a final state that is almost axisymmetric. The short-to-long axis ratio also increases, from its initial value of ~0.5 to ~0.9 near the center, and ~0.6 at the half-mass radius; the elongation of the model ceases to change once axisymmetry is reached, a fact which argues in favor of the evolution being driven by the box orbits.

Although the final shape of the model is nearly the same for the three different values of M_{BH} tested here, the evolution time was found to depend strongly on the black hole mass. When $M_{BH} = 0.003M_{gal}$, axisymmetry was not quite reached by the end of the integration, at roughly 55 half-mass orbital periods. But when M_{BH} was increased to $0.03M_{gal}$, the model evolved in shape on approximately the same timescale that the black hole grew. Experiments with smaller values of t_{grow} revealed that the response time of the galaxy to the black hole was essentially instantaneous — i.e., of order the local orbital period — when M_{BH} exceeded ~2% of the galaxy mass. Such rapid evolution is just what would be expected based on the test particle integrations in FIG. 5. We conjecture that a transition to global stochasticity occurs at $M_{BH} \approx 0.02M_{gal}$, causing the boxlike orbits to lose their characteristic shapes in a single orbital period. The galaxy responds by rapidly becoming axisymmetric.

This rapid evolution toward axisymmetry may constitute a negative feedback mechanism that determines the maximum mass of nuclear black holes [19]. In one widely-discussed scheme [32], nuclear black holes grow by capturing stars or gas clouds on box orbits as they pass near the galactic center. But as the mass of the black hole approaches ~2% of the galaxy's mass, the galaxy would evolve rapidly to axisymmetry, thus eliminating the box orbits and cutting off the supply of stars or gas to the black hole. Black holes could then never accrete more than ~2% of their host galaxy's mass unless they somehow managed to grow on a timescale much shorter than a galaxy crossing time. This hypothesis is consistent with what is known so far about the masses of nuclear black holes, the largest of which have masses of order 1% the stellar mass of their parent galaxies [8]. The current record-holder is NGC 3115 [33], which has $M_{BH}/M_{gal} \approx 2.5\%$, nicely consistent with the prediction.

Although the critical black hole mass found here was derived from a single triaxial model with a particular shape, roughly the same mass may define the transition to global stochasticity in models with very different geometries. For instance, Norman, Sellwood and Hasan [34] found that 2-D models of rotating barred spiral galaxies exhibited strong evolution toward axisymmetry when M_{BH}/M_{gal} exceeded ~5%. But in 3-D simulations of similar barred systems [35], the bar was found to be much more fragile, evolving strongly once the black hole mass fraction exceeded ~1%.

ACKNOWLEDGMENT

The *N*-body work described here was carried out in collaboration with G. Quinlan.

REFERENCES

1. CHANDRASEKHAR, S. 1942. Principles of Stellar Dynamics. University of Chicago Press. Chicago, IL.

2. HUNTER, C. 1995. Ann. NY Acad. Sci. **751:** 76–92.
3. KUZMIN, G.G. 1956. Astr. Zh. **33:** 27–35.
4. KUZMIN, G.G. 1973. *In* Dynamics of Galaxies and Clusters. T.B. Omarov, Ed.: 71–75. Akad. Nauk. Kaz. SSR. Alma Ata.
5. DE ZEEUW, PT. 1985. Mon. Not. R. Astron. Soc. **216:** 273–334.
6. GEBHARDT *et al.* 1996. Astron. J. **112:** 105–113.
7. MERRITT, D. & T. FRIDMAN. 1995. *In* Fresh Views of Elliptical Galaxies, A.S.P. Conf. Ser. Vol. 86. A. Buzzoni, A. Renzine & A. Serrano, Eds.: 13–22. Astronomical Society of the Pacific. Provo, Utah.
8. KORMENDY, J. & D. O. RICHSTONE. 1995. Ann. Rev. Astron. Astrophys. **33:** 581–624.
9. GOODMAN, J. & M. SCHWARZSCHILD. 1981. Astrophys. J. **245:** 1087–1093.
10. FRIDMAN, T. & D. MERRITT. 1997. Astron. J. **114:** 1479–1487.
11. MIRALDA-ESCUDE, J. & M. SCHWARZSCHILD. 1989. Astrophys. J. **339:** 752–762.
12. GERHARD, 0. & J. J. BINNEY. 1985. Mon. Not. R. Astron. Soc. **216:** 467–502.
13. SCHWARZSCHILD, M. 1979. Astrophys. J. **232:** 236–247.
14. SCHWARZSCHILD, M. 1982. Astrophys. J. **263:** 599–610.
15. STATLER, T. 1987. Astrophys. J. **321:** 113–152.
16. MERRITT, D. & T. FRIDMAN. 1996. Astrophys. J. **460:** 136–162.
17. MERRITT, D. 1997. Astrophys. J. **486:** 102–109.
18. VALLURI, M. & D. MERRITT. 1998. Astrophys. J. Submitted.
19. MERRITT, D. & G. QUINLAN. 1998. Astrophys. J. **498:** in press.
20. MERRITT, D. & M. VALLURI. 1996. Astrophys. J. **471:** 82–105.
21. LASKAR, J., C. FROESCHLE & A. CELLETTI. 1992. Physica D. **56:** 253–269.
22. LASKAR, J. 1996. *In* NATO Advanced Study Institute, Hamiltonian Systems with Three or More Degrees of Freedom. C. Simo & A. Delshams, Eds. Kluwer.
23. PAPAPHILIPPOU, Y. & J. LASKAR. 1996. Astron. Astrophys. **307:** 427–449.
24. PAPAPHILIPPOU, Y. & J. LASKAR. 1997. Astron. Astrophys. **329:** 451–481.
25. DEHNEN, W. 1993. Mon. Not. Royal Astron. Soc. **265:** 250–256.
26. LASKAR, J. 1993. Physica D. **67:** 257–281.
27. SCHWARZSCHILD, M. 1993. Astrophys. J. **409:** 563–577.
28. SELLWOOD, J.A. 1997. *In* Computational Astrophysics, A.S.P. Conf. Ser. Vol. 123. D.A. Clarke & M.J. West, Eds.: 215–220. Astronomical Society of the Pacific. Provo, Utah.
29. AGUILAR, L. & D. MERRITT. 1990. Astrophys. J. **354:** 33–51.
30. PEEBLES, P.J.E. 1972. Gen. Rel. Grav. **3:** 63–72.
31. NORMAN, C.A., A. MAY & T.S. VAN ALBADA. 1985. Astrophys. J. **296:** 20–34.
32. NORMAN, C.A. & J. SILK. 1993. Astrophys. J. **266:** 502–515.
33. KORMENDY, J., R. BENDER & S. TREMAINE. 1996. Astrophys. J. **459:** L57–L60.
34. NORMAN, C. A., J. A. SELLWOOD & H. HASAN. 1996. Astrophys. J. **462:** 114–124.
35. SELLWOOD, J. A. & E. MOORE. In preparation.

Gravity and Large-Scale Structure: Observational Evidence[a]

J. N. FRY[b]

Department of Physics, University of Florida, Gainesville, Florida 32611-8440

ABSTRACT: In our present understanding, cosmological structure on large scales is formed by the action of gravity amplifying small initial fluctuations. On large scales, where fluctuations are weak and perturbation theory applies, that presumption can be tested against observations. In particular, the anisotropic shapes of structures formed by gravity predicts a characteristic behavior of the bispectrum, or third moment of density in Fourier space, that can be used as a signature to separate effects of gravity from more local, nongravitational forces that may play a role in galaxy formation.

1. INTRODUCTION

The current picture of the origin of structure on cosmologically large scales begins with small initial density fluctuations, perhaps imprinted in an early epoch of inflation that then grow by gravitational instability into the observed clustering hierarchy, including galaxies, groups, structures, and superclusters. That the initial fluctuations are small and most likely Gaussian is supported by observations of temperature fluctuations in the cosmic microwave background radiation [36], but that structure is in fact shaped by gravity is more difficult to demonstrate.

Even for a purely gravitational system the evolution of density is a nonlinear problem, and only cases with special symmetry can be solved exactly. The process of galaxy formation, while not well understood, presumably brings in nongravitational forces, collisions, shocks, transport, dissipation, etc., and it is not at all clear how much the present universe is related to initial conditions. However, on large scales, the evolution of weak fluctuations can be studied in perturbation theory, and, to the extent that the process of galaxy formation can be considered local, this leads to specific predictions for the behavior of galaxy correlation functions in which the effects of gravity and galaxy formation can be identified separately. This appears in a straightforward and graphic way in the dependence of the third moment of the galaxy distribution in the transform domain, or the bispectrum, on the shape of the triangular configuration of wavenumbers.

This paper sketches how this comes about. Section 2 recaps the theory of the evolution of density fluctuations in an expanding universe in perturbation theory and shows how mode coupling produces a third moment even for an initially Gaussian distribution. In Section 3, I include the possibility of nongravitational forces acting in the formation of galaxies and compute the effects on galaxy correlations of an arbitrary nonlinear but local bias. Section 4 reports some preliminary observational re-

[a]This research was supported in part by NASA grant NAG5-2835.
[b]E-mail: fry@phys.ufl.edu.

sults. In Section 5, I outline further considerations that may have an effect on the dependence on configuration shape, including projection in angular catalogs, peculiar velocities in redshift space, nonlinear evolution, and non-Gaussian initial conditions. Section 6 summarizes the present state of the subject and future prospects.

2. DYNAMICAL EVOLUTION OF STRUCTURE IN PERTURBATION THEORY

A. Dynamics

The present universe is almost homogeneous on large scales, with a matter dominated average density that dilutes with time as the universe expands as $\bar{\rho} \sim 1/a^3(t)$, where $a(t)$ is the usual cosmological scale factor (objects at fixed comoving separation $\boldsymbol{x}$ have proper separation $\boldsymbol{r} = a\boldsymbol{x}$) [29]. In Einstein or Newtonian gravity the scale factor obeys

$$\left(\frac{\dot{a}}{a}\right)^2 = \frac{8\pi G\bar{\rho}}{3} - \frac{k}{a^2}. \quad (1)$$

For the case of critical density ($\Omega = 1$), zero curvature ($k = 0$), and zero cosmological constant ($\Lambda = 0$), the scale factor depends on time as $a(t) \sim t^{2/3}$, and time and density are related as $6\pi G\bar{\rho}\, t^2 = 1$.

Departures from homogeneity can then be described by the density contrast δ, $\rho(\boldsymbol{x}, t) = \bar{\rho}\,[\,1 + \delta(\boldsymbol{x},t)]$, and the peculiar velocity $\boldsymbol{v}$ relative to the expanding background. From moments of the Vlasov equation in the Newtonian limit [29], [12], the evolution of density is governed by the equation of continuity,

$$\frac{\partial\delta}{\partial t} + \frac{1}{a}\nabla\cdot[(1+\delta)\boldsymbol{v}] = 0\,, \quad (2)$$

and the Euler equation in an expanding background,

$$\frac{1}{a}\frac{\partial}{\partial t}(a\boldsymbol{v}) + \frac{1}{a}(\boldsymbol{v}\cdot\nabla)\boldsymbol{v} + \frac{1}{a}\nabla\phi = 0\,, \quad (3)$$

where the gravitational potential obeys the Poisson equation

$$\nabla^2\phi = 4\pi G\bar{\rho}\, a^2\delta. \quad (4)$$

This last has the well-known solution $\phi = -4\pi G\bar{\rho}\, a^2\Delta$, where

$$\Delta(\boldsymbol{x}) = \frac{1}{4\pi}\int d^3x'\,\frac{\delta(\boldsymbol{x}')}{|\boldsymbol{x}-\boldsymbol{x}'|} \quad (5)$$

is the solution to $\nabla^2\,\Delta = -\delta$.

A time derivative of the equation of continuity and the divergence of the Euler equation can be combined to obtain the standard master equation for the evolution of density fluctuations [29], [12],

$$\frac{\partial^2\delta}{\partial t^2} + \frac{2\dot{a}}{a}\frac{\partial\delta}{\partial t} - 4\pi G\bar{\rho}\delta = 4\pi G\bar{\rho}\delta^2 + \frac{1}{a^2}\nabla\phi\cdot\nabla\delta + \frac{1}{a^2}\nabla_i\nabla_j[(1+\delta)v^i v^j] \; . \tag{6}$$

B. Linear Perturbations

For a system with no particular symmetry, Eq. (6) in general can only be solved numerically or modelled in large numerical simulations. However, when δ is small we can attempt a systematic perturbative expansion. The left-hand side, containing only linear terms, has no spatial derivatives. Thus, to linear order, initial fluctuations grow by an overall factor,

$$\delta(\boldsymbol{x}, t) = A(t)\delta(\mathrm{x}, t_0) = A(t)\delta_0(\boldsymbol{x}). \tag{7}$$

For a matter-dominated, critical density universe, this amplitude factor behaves as $A(t) \sim t^{2/3} \sim a(t)$. Note that at linear order this amplification by a universal scale factor means that initially Gaussian fluctuations remain Gaussian.

From the equation of continuity, the peculiar velocity is then

$$v^i = HF(\Omega)\nabla^i\Delta, \tag{8}$$

where H is the Hubble constant ($H = \dot{a}/a$ in physical coordinates; $H = \dot{a}$ in comoving coordinates), and $F(\Omega) = (\dot{A}/A)/(\dot{a}/a) = d\log A/d\log a$ [29] is the ratio of the rate of growth of fluctuations to the rate of cosmological expansion. For Ω (the fraction of critical density) near 1, $F \approx \Omega^{4/7}$ [10]. Comparing observed peculiar velocities with the distribution of galaxies thus can provide an estimate of the density parameter Ω, provided galaxies fairly trace mass.

C. Higher-order Perturbations

On large scales it is convenient to work in the Fourier domain,

$$\tilde{\delta}(\boldsymbol{k}) = \int d^3x\,\delta(\boldsymbol{x})e^{-i\boldsymbol{k}\cdot\boldsymbol{x}}, \quad \delta(\boldsymbol{x}) = \int\frac{d^3k}{(2\pi)^3}\tilde{\delta}(\boldsymbol{k})e^{ik\cdot x}\,. \tag{9}$$

As in any nonlinear theory, the higher-order terms in Eq. (6) lead to a mixing of modes, and in general the Fourier amplitude $\tilde{\delta}(\boldsymbol{k})$ is a sum of terms, $\tilde{\delta}(\boldsymbol{k}) = \tilde{\delta}^{(1)}(\boldsymbol{k}) + \tilde{\delta}^{(2)}(\boldsymbol{k}) + \tilde{\delta}^{(3)}(\boldsymbol{k}) + \ldots$, where $\tilde{\delta}^{(n)} \sim A^n\tilde{\delta}_0{}^n$. The nth-order term in full is a convolution [18], [23],

$$\tilde{\delta}^{(n)}(\boldsymbol{k}) = \int\frac{d^3k_1'}{(2\pi)^3}\cdots\frac{d^3k_n'}{(2\pi)^3}\,[(2\pi)^3\delta_D(\Sigma\boldsymbol{k}_i' - \boldsymbol{k})]\,G_n(\boldsymbol{k}_1', \ldots, \boldsymbol{k}_n')\;\tilde{\delta}(\boldsymbol{k}_1')\ldots\tilde{\delta}(\boldsymbol{k}_n'), \tag{10}$$

where $\tilde{\delta}(\boldsymbol{k}_i')$ on the right-hand side is the linearly evolved initial fluctuation $A(t)\,\tilde{\delta}_0(\boldsymbol{k}_i')$. The kernel G_n is a symmetric, scalar function, homogeneous of degree zero; for $n = 2$ [9]

$$G_2(\boldsymbol{k}_1, \boldsymbol{k}_2) = \frac{5}{7} + \frac{1}{2}\hat{\boldsymbol{k}}_1 \cdot \hat{\boldsymbol{k}}_2\left(\frac{k_1}{k_2} + \frac{k_2}{k_1}\right) + \frac{2}{7}(\hat{\boldsymbol{k}}_1 \cdot \hat{\boldsymbol{k}}_2)^2 . \tag{11}$$

In an open universe the factors $\frac{5}{7}$ and $\frac{2}{7}$ become $\frac{1}{2} \pm \kappa$, where κ depends only weakly on Ω ; for Ω near 1,

$$\kappa \sim \frac{3}{14}\ \Omega^{-2/63},$$

while $\kappa \to \frac{1}{4}$ as $\Omega \to 0$ [3], [24]. Higher-order G_n are obtained by an algebraic recursion [18], [23], [34].

D. Moments

In general, $\delta(\boldsymbol{x})$ is a fluctuating field with no significance attached to its value at a particular point, but moments or correlation functions reflect the underlying homogeneity and isotropy. The first few of these are the two-, three-, and four-point functions,

$$\langle \delta(\boldsymbol{x}_1)\ \delta(\boldsymbol{x}_2) \rangle = \xi(\boldsymbol{x}_{12}) = \xi_{12}, \tag{13}$$

$$\langle \delta(\boldsymbol{x}_1)\ \delta(\boldsymbol{x}_2)\ \delta(\boldsymbol{x}_3) \rangle = \zeta(\boldsymbol{x}_1, \boldsymbol{x}_2, \boldsymbol{x}_3), \tag{14}$$

$$\langle \delta(\boldsymbol{x}_1)\ \delta(\boldsymbol{x}_2)\ \delta(\boldsymbol{x}_3)\ \delta(\boldsymbol{x}_4) \rangle = \xi_{12}\ \xi_{34} + \xi_{13}\ \xi_{24} + \xi_{23}\ \xi_{14} + \eta_{1234}, \tag{15}$$

where $\boldsymbol{x}_{12} = |\boldsymbol{x}_1 - \boldsymbol{x}_2|$; the three- and four-point functions ζ and η also depend only on relative separations and not on absolute position or orientation. Starting with the four-point function we must include corrections for unconnected terms to obtain the connected or irreducible moment. In general, we write the nth connected moment as $\xi_n = \langle \delta_1 \ldots \delta_n \rangle_c$. In observations it is found that for configurations of similar shape that differ only by an overall scale r, $\xi \sim r^{-\gamma}$, $\zeta \sim r^{-2\gamma}$ [19], and $\eta \sim r^{-3\gamma}$ [15]. This has led to speculation that ξ_n scales as ξ^{n-1}. For Gaussian initial conditions with a power-law initial correlation function, this is exactly the behavior of the first nonvanishing contribution to ξ_n in perturbation theory. Even though the initial fluctuations are expected to be Gaussian, nonlinearity induces nonvanishing higher-order moments to all orders. For example, for the third moment we have

$$\langle \delta^3 \rangle = \langle [\delta^{(1)} + \delta^{(2)} + \ldots\ ]^3 \rangle = \langle [\delta^{(1)}]^3 \rangle + 3\ \langle [\delta^{(1)}]^2\ \delta^{(2)} \rangle + \ldots . \tag{16}$$

For Gaussian δ the first term vanishes, but the second does not. In general, $\delta^{(n-1)}$ is required to compute the first nonvanishing contribution ξ_n. It is possible to write down a functional form that expresses ξ_n as a sum of tree graphs, with n points connected by $n - 1$ links [9], [12], much as in quantum field theory; this is often denoted the "tree-level" result. Irreducible moments of the density contrast at a single point can be written in terms of the variance $\langle \delta^2 \rangle = \xi$ as $\xi_n = \langle \delta^n \rangle_c = S_n\, \xi^{n-1}$, where the S_n are dimensionless hierarchical amplitudes. The first few of these are $S_3 = \frac{34}{7}$ [29],

[9], and $S_4 = \frac{60712}{1323}$ [9]. A vertex generating function approach allows calculation of the general coefficient S_n [2].

On large scales, it is again convenient to examine the full multipoint dependence in Fourier space. The first few moments of $\tilde{\delta}(k)$ are the power spectrum and the bispectrum,

$$\langle \tilde{\delta}(\boldsymbol{k}_1)\tilde{\delta}(\boldsymbol{k}_2)\rangle = [(2\pi)^3\delta_D(\boldsymbol{k}_1+\boldsymbol{k}_2)]P(k), \tag{17}$$

$$\langle \tilde{\delta}(\boldsymbol{k}_1)\tilde{\delta}(\boldsymbol{k}_2)\tilde{\delta}(\boldsymbol{k}_3)\rangle = [(2\pi)^3\delta_D(\boldsymbol{k}_1+\boldsymbol{k}_2+\boldsymbol{k}_3)]B(\boldsymbol{k}_1,\boldsymbol{k}_2,\boldsymbol{k}_3). \tag{18}$$

The "momentum-conserving" Dirac δ-function, restricting the bispectrum B to closed triangles $\Sigma k_i = 0$, is a consequence of homogeneity. From above, perturbation theory at leading nonvanishing order, or "tree" level, gives

$$B(\boldsymbol{k}_1,\boldsymbol{k}_2,\boldsymbol{k}_3) = Q_{12}P(k_1)P(k_2) + Q_{13}P(k_1)P(k_3) + Q_{23}P(k_2)P(k_3), \tag{19}$$

where for $\Omega = 1$

$$Q_{ij} = \frac{10}{7} + \hat{\boldsymbol{k}}_i\cdot\hat{\boldsymbol{k}}_j\left(\frac{k_i}{k_j}+\frac{k_j}{k_i}\right) + \frac{4}{7}\,(\hat{\boldsymbol{k}}_i\cdot\hat{\boldsymbol{k}}_j)^2. \tag{20}$$

As suggested from observations, it is useful to define the hierarchical amplitude

$$Q_{123} = \frac{B_{123}}{P_1P_2+P_1P_3+P_2P_3}. \tag{21}$$

With this normalization, the growth factor $A(t)$ divides out, and Q is time independent. Further, while Q does depend on the power spectrum, if $P(k)$ is a power law, $P \sim k^n$, then Q does not depend on overall scale but only on angles and ratios of sides, the shape of the triangle configuration of wave vectors. Q is also remarkably insensitive to the density parameter Ω, over the whole range from $\Omega = 1$ to $\Omega = 0$. FIGURE 1 shows Q for configurations with $k_1 = 1$, $k_2 = 1/2$, separated by θ for four different power spectrum indices, $n = -2, -1, 0$, and $+1$. The preference for linear configurations for $n \leq 0$ reflects the anisotropy of gravitational collapse, to sheets or filaments. The behavior of the three-point amplitude Q is a specific and robust prediction of the action of gravity.

3. BIAS IN GALAXY FORMATION

The result obtained for Q has no free parameters once the power spectrum is known, and thus can be taken as a signature of the action of gravity. However, the formation of galaxies, which are what we observe, may bring in other forces. The process of galaxy formation is not well understood, but as it may involve hydrodynamic shocks, dissipation, transport, it is quite possible that it does not occur with equal efficiency per gram of hydrogen everywhere; i.e., it is possible that there is bias in galaxy formation. It is clear that not all galaxies trace mass equally: a depen-

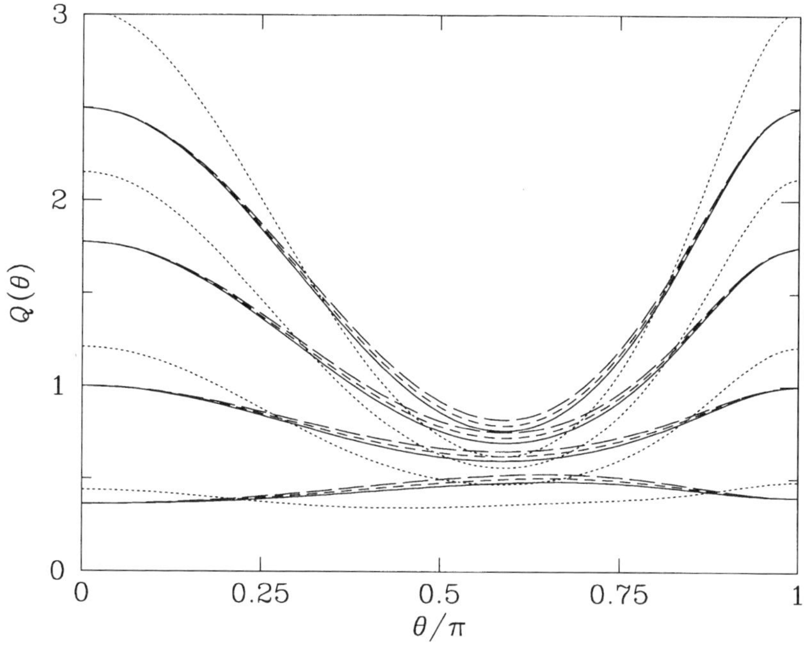

FIGURE 1. Dependence of the three-point amplitude Q on configuration shape expected from perturbation theory for power spectrum $P(k) \sim k^n$ with $n = -2$, $n = -1$, $n = 0$, and $n = +1$ (groups from top to bottom). Solid lines show $\Omega = 1$, long-dashed lines show $\Omega = 0.1$, and short-dash lines $\Omega \to 0$. Dotted lines show expected behavior in redshift space for $\Omega = 1$.

dence of morphology on environment [8], [30] leads to stronger clustering of elliptical than of spiral galaxies [5]; clustering strength increases for galaxies with higher surface brightness [6] and for intrinsically brighter galaxies [21]; and optically selected galaxies are more strongly clustered than infrared selected galaxies [37], [31].

Many toy models in which galaxies form at preferred locations such as peaks in the mass distribution [26], [1], have been found to give a galaxy density that has an enhanced contrast relative to the mass density, $\delta_g = b\,\delta$ where $\delta_g = [n(\boldsymbol{x}) - \bar{n}]/\bar{n}$ and $\delta = [\rho(\boldsymbol{x}) - \bar{\rho}]/\bar{\rho}$. In fact, this is the leading behavior for an arbitrary local bias. If the distribution of matter physically determines the resulting distribution of galaxies, then mathematically the galaxy density is a functional of the matter density, $n_g(\boldsymbol{x}) = F[\rho(\boldsymbol{x})]$. If the range of influence is short, so that only local conditions apply, we have a so-called local bias [26], [4], $n_g(\boldsymbol{x}) = f(\rho(\boldsymbol{x}))$. Expressed in terms of the density contrast, we have

$$\delta_g = \sum \frac{1}{k!} b_k \delta^k . \qquad (22)$$

The coefficients b_k parametrize the bias function; the usual linear bias arises from the leading term, $b = b_1$.

Galaxy correlations then depend on the bias parameters b_k as well as the underlying density correlations [13]. To leading order in perturbation theory, density fluctuations have the same shape as initial conditions and the galaxy two-point function and power spectrum are simple amplifications of the corresponding matter functions by the linear coefficient $b = b_1$, $\xi_g = b^2\xi$, $P_g = b^2 P$. Expressed in terms of δ_g, peculiar velocities are as in Eq. (8), with $F(\Omega)$ replaced with a parameter $\beta = \Omega^{4/7}/b$. The bispectrum to consistent order includes the second-order nonlinear bias coefficient b_2 [13], [25],

$$Q_g = \frac{1}{b}Q_{123} + \frac{b_2}{b^2}. \tag{23}$$

Since it is usually the case that linear bias enhances galaxy correlations, $b > 1$, the effect of bias is to weaken the configuration dependence of the three-point amplitude.

4. GALAXY OBSERVATIONS

We can apply these considerations to galaxy observations. The power spectrum and bispectrum were computed by Fry and Seldner [17] for a square area around the north galactic pole in the Lick Observatory catalog of Shane and Wirtanen [35]. The Lick catalog has characteristic depth $D^* = 210 h^{-1}$ Mpc [19] (h is Hubble's constant in units of 100 km s^{-1}Mpc^{-1}), while the scale of nonlinearity in galaxy clustering is roughly 8 h^{-1}Mpc, so that on scales beyond 4% of the depth of the catalog perturbation theory should apply.

The power spectrum in the catalog is seen to go as $P(k) \sim k^n$ with index $n \approx -1.4$ for wavenumbers $5 \le k \le 30$ [17], [11] (k in units of "waves per box;" physical wavelengths are $\lambda = 260\ h^{-1}$ Mpc$/k$). FIGURE 2 shows $Q(\theta)$ averaged for triangles with legs in the same $5 \le k_i \le 30$ range, with two legs having ratio $0.4 < k_2/k_1 < 0.6$. Results from the projected angular data have been scaled to give Q for the true, three-dimensional distribution by an overall multiplicative correction factor of roughly 1.3 [19].

A small but significant departure from Q = constant with the form expected from gravitational instability is apparent. Fit to a constant Q, the data in FIGURE 2 have a reduced $\chi^2 = 1.10$ (for 14 degrees of freedom), while a fit to the form in Eq. (23) for $n = -1.4$ gives $\chi^2 = 0.41$ (13 degrees of freedom), while using all data for the same range of scales, not just that in the figure, $\chi^2/241 = 1.12$ for Q = constant and $\chi^2/240 = 1.04$ for Eq. (23) [11]. Most important, the same values for the fit parameters result from the restricted and complete data samples, $b = 3.5 \pm 1.1$, $b_2/b^2 = 1.2 \pm 0.1$ (data in figure) and $b = 3.0 \pm 0.65$, $b_2/b^2 = 1.1 \pm 0.1$ (all data) [11]. This value for b is somewhat higher than from dynamical methods [38], [7], but not by much more than one standard deviation. The result is possibly affected by one or more of the effects listed in the next section.

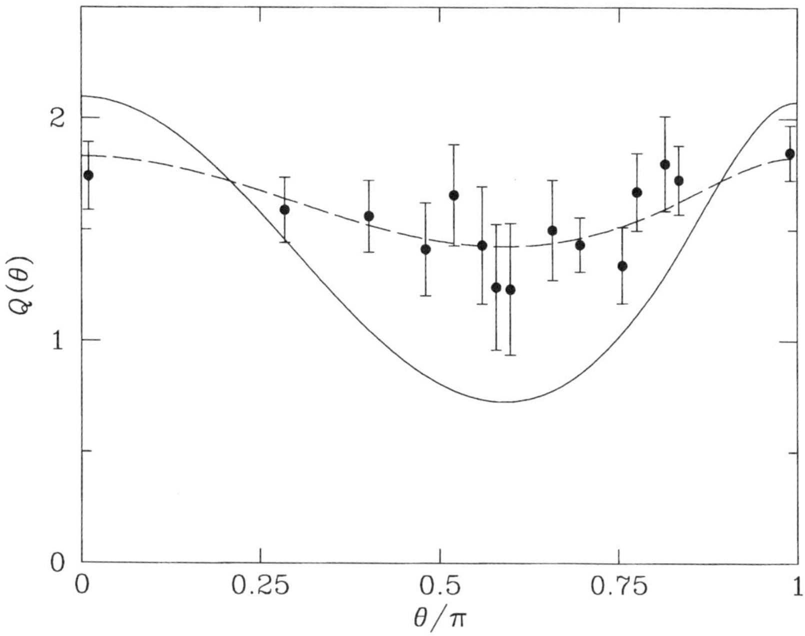

FIGURE 2. Observed $Q(\theta)$ from the Lick Observatory catalog. The solid line is the expected result from gravitational instability without bias for the observed power spectrum index $n = -1.41$. The dashed line is fit to Eq. (23) with $b = 3.5$ and $b_2/b^2 = 1.2$.

5. OTHER CONSIDERATIONS

Other observational and evolution effects can modify the dependence of Q on configuration shape and mimic the effects of bias, and thus must be accounted for in the analysis. Some of these, including radial projection in angular catalogs, peculiar velocity distortions in redshift space, nonlinear evolution of fluctuations, and non-Gaussian initial conditions, are briefly summarized here.

A. Projection

At this time, the largest galaxy catalogs contain only angular positions, projected over distance. For a Cartesian projection with a selection function $F(z)$ that reflects the probability that an object at distance z is included, the power spectrum and bispectrum P_p and B_p of the projected distribution $\delta_p(\boldsymbol{x}) = \int dz\, F(z)\, \delta(\boldsymbol{x}, z)$ are related to the intrinsic P and B smeared over $\tilde{F}$, the Fourier transform of the selection function [11],

$$P_p(k_p) = \int \frac{dk'}{2\pi} \left|\tilde{F}(k')\right|^2 P((k_p^2 + k'^2)^{1/2}), \tag{24}$$

$$B_p(k_{p,i}) = \int \frac{dk_1'}{2\pi}\frac{dk_2'}{2\pi}\tilde{F}(k_1')\tilde{F}(k_2')\tilde{F}(k_3')B(\boldsymbol{k}_{p,i}, k_i'), \tag{25}$$

where $k_3' = -k_1' - k_2'$. Typically $F(z)$ is appreciable only to some characteristic depth D^*, and by the uncertainty principle $\tilde{F}(k')$ has characteristic extent $1/D^*$. For small wavelengths, $kD^* > 1$, the main contribution to the integrals is then for $k' \ll k$, and the projected power spectrum and bispectrum can be taken of the integral, evaluated in the plane $k' = 0$. The remaining integrals are then simple numerical factors equivalent to those studied in detail for the direct correlation functions [19], and up to this factor the projected spectra preserve all the shape information in B. However, work in progress [39] suggests that the mixing of scales in angular instead of Cartesian projection puts a lower limit on the amount of smearing, and thus may bring the inferred value of b closer to that from other sources.

B. Redshift Distortions

One might hope to avoid the problems of projection by using a full three-dimensional catalog. However, radial distances are most often assigned from redshifts, leading to distortions arising from peculiar velocities [27]. The transformation from coordinate to redshift distances modulates the density,

$$1 + \delta_z = \frac{1+\delta}{1 + H^{-1}dv_z/dz}, \tag{26}$$

to linear order in the Fourier domain gives an enhancement of the galaxy fluctuation amplitude in the radial direction,

$$\tilde{\delta}_z'(\boldsymbol{k}) = 1 + \left(\frac{F(\Omega)}{b}(\hat{\boldsymbol{k}}\cdot\hat{\mathbf{z}})^2\right)\hat{\delta}(\boldsymbol{k}). \tag{27}$$

Averaged over orientations, the power becomes

$$P_z(k) = \left(1 + \frac{2}{3}\frac{F}{b} + \frac{1}{5}\frac{F^2}{b^2}\right)P(k). \tag{28}$$

At higher order the calculation is long but straightforward [22]; the net effect is to increase somewhat the enhancement for colinear configurations, as shown in the dotted lines in FIG. 1 for $\Omega =$. For $\Omega \rightarrow 0$, velocities and thus velocity distortions vanish.

C. Nonlinear Evolution

Perturbation theory is valid only on scales where rms fluctuations are small, or for scales $k < k_{nl}$, where the nonlinear scale k_{nl} is defined by

$$\int_0^{k_{nl}} \frac{d^3k}{(2\pi)^3}P(k) = 1. \tag{29}$$

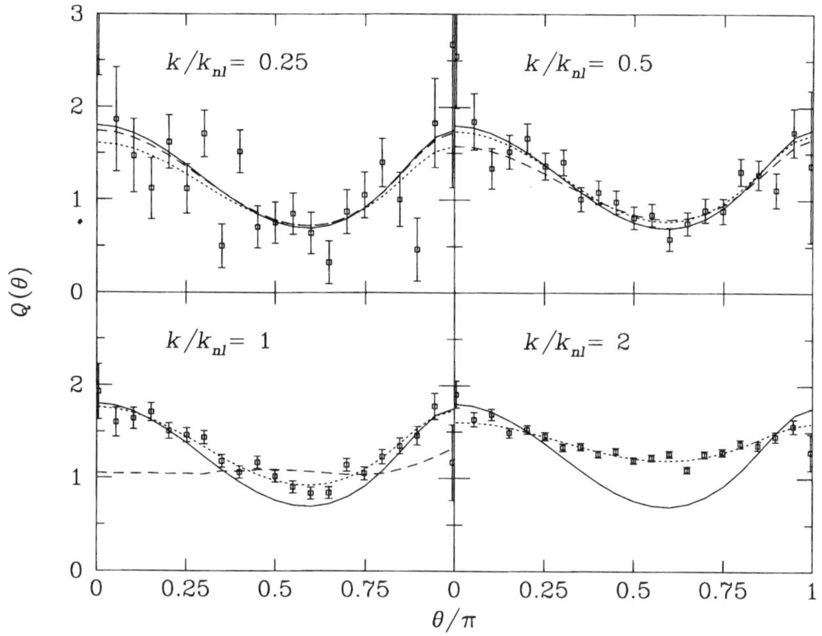

FIGURE 3. Nonlinear evolution of $Q(\theta)$ from numerical simulations and corrections from higher-order terms in perturbation theory at four scales with increasing fluctuation amplitude. The solid line is the leading result from perturbation theory for power spectrum index $n = -1$; the dashed line includes the "one-loop" correction. Points and errors are from numerical simulations, and the dotted line is the best fit of the data to a biased tree-level prediction. At $k = k_{nl}$, interpreting the evolved result as bias applied to the tree-level perturbation theory form gives an apparent bias parameter $b = 1.3$.

The onset of effects of nonlinear evolution can be studied numerically or by computing higher-order terms in perturbation theory [32]–[34]. As the amplitude of fluctuations grows, next to leading order or "one-loop" corrections agree well with the first departures from the tree-level result seen in numerical simulations [14], as shown in FIG. 3 for $n = -1$, for four scales characterized as a fraction of k_{nl}. The results are much as one would expect. On the largest scales, the leading tree-level term is adequate, and the next leading term describes well the first correction. However, approaching the nonlinear scale k_{nl} (where perturbation theory is expected to break down) the numerical evolution progresses more slowly than the one-loop correction would suggest, while even further into the nonlinear regime the numerical results show a much flatter dependence on configuration shape. Nonlinear evolution does not have much effect for $k < k_{nl}/2$.

D. Non-Gaussian Initial Conditions

Finally, there is also the possibly non-Gaussian initial conditions. At a single point, the second moment of density for non-Gaussian initial conditions depends on the initial three-point function [16]

$$\xi(0) = \langle\delta(\boldsymbol{x})^2\rangle = \xi_0(0) + \frac{13}{21}\zeta_0(0) + \frac{4}{7}I[\zeta_0], \tag{30}$$

and third moment involves the initial three- and four-point functions

$$\zeta(0) = \langle\delta(\boldsymbol{x})^3\rangle = \zeta_0(0) + \frac{34}{7}\xi_0^2(0) + \frac{10}{7}\eta_0(0) + \frac{6}{7}I[\eta_0], \tag{31}$$

$$S_3 = \frac{S_{3,0}}{a(t)} + \frac{34}{7} + \frac{10}{7}\frac{\eta_0}{\xi_0^2} + \frac{6}{7}\frac{I[\eta_0]}{\xi_0^2} - \frac{26}{21}\frac{\zeta_0^2}{\xi_0^3} - \frac{8}{7}\frac{\zeta_0 I[\zeta_0]}{\xi_0^3}, \tag{32}$$

where the integral $I[\zeta]$ is

$$I[\zeta] = \int\frac{d^3x'}{4\pi}\frac{d^3x''}{4\pi}\frac{6P_2(\hat{\boldsymbol{x}}'\cdot\hat{\boldsymbol{x}}'')}{x'^3x''^3}\zeta(0, x', x'')\cdot \tag{33}$$

[$P_2(\mu) = \frac{3}{2}\mu^2 - \frac{1}{2}$ is the second Legendre polynomial] and $I[\eta]$ is the same with $\eta(\boldsymbol{x}', \boldsymbol{x}'', 0, 0)$.

The full three-point function or bispectrum will depend on the properties of the specific non-Gaussian model but will involve the initial third and fourth moments in analogous ways. At this time, there is no reason we need to invoke non-Gaussian initial conditions.

6. CONCLUSIONS

We have seen that the evolution of large-scale cosmological structure in gravitational instability predicts a specific dependence of the three-point function or bispectrum on configuration shape, which can be taken as a signature of structure formation by gravity. Observations of the bispectrum of galaxies show a bare hint of a dependence on configuration shape, with a preference for colinear configurations as required by gravity. The observed effect is weaker than gravity itself predicts, probably reflecting a product of nonlinear evolution, projection, and a modest nonlinear bias in galaxy formation. There is at present no need to invoke non-Gaussian initial conditions, but future observations that show a configuration shape dependence that is not consistent with gravity could require either this possibility or a modification of the simple local bias model.

Future surveys will provide a much stronger picture. The effects of projection can be avoided by working in redshift space. The effects of nonlinear evolution can be minimized by looking at larger scales, but observations are increasingly difficult, with errors expected to grow as $(k_{nl}/k)^{3+n}$ on large scales (small k). The optimum survey would be a densely sampled redshift survey over a volume that is many times larger the scale of galaxy clustering. This is a description of the planned Sloan Digital Sky Survey [20], [28]. Perhaps soon, by measuring galaxy statistics, it will be possible to determine whether large scale structure is indeed shaped by gravity.

REFERENCES

1. BARDEEN, J. M., J.R. BOND, N. KAISER & A.S. SZALAY. 1986. The statistics of peaks of Gaussian random fields. Astrophys. J. **304:** 15–61.
2. BERNARDEAU, F. 1992. The gravity-induced quasi-Gaussian correlation hierarchy. Astrophys. J. **392:** 1.
3. BOUCHET, F.R., R. JUSZKIEWICZ, S. COLOMBI & R. PELLAT. 1992. Weakly nonlinear gravitational instability for arbitrary Ω. Astrophys. J. Lett. **394:** L5–L8.
4. COLES, P. 1993. Galaxy formation with a local bias. Mon. Not. R. Astron. Soc. **262:** 1065–1075.
5. DAVIS, M. & M. J. GELLER. 1976. Galaxy correlations as a function of morphological type. Astrophys. J. **208:** 13–19.
6. DAVIS, M. & S. DJORGOVSKI. 1985. Galaxy clustering as a function of surface brightness. Astrophys. J. **299:** 15–20.
7. DEKEL, A., E. BERTSCHINGER, A. YAHIL, M. A. STRAUSS, M. DAVIS & J. P. HUCHRA. 1993. *IRAS* galaxies vs. POTENT mass: Density fields, biasing, and Ω. Astrophys. J. **421:** 1–21.
8. DRESSLER, A. 1980. Galaxy morphlogy in rich clusters: Implications for the origin and evolution of galaxies. Astrophys. J. **236:** 351–365.
9. FRY, J.N. 1984. The galaxy correlation hierarchy in pertubation theory. Astrophys. J. **279:** 499–510.
10. FRY, J.N. 1985. Dynamical measures of density in exotic cosmologies. Phys. Lett. **158B:** 211–214.
11. FRY, J.N. 1994. Gravity, bias, and the galaxy three-point correlation function. Phys. Rev. Lett. **73:** 215–218.
12. FRY, J.N. 1998. Physics Reports. Submitted.
13. FRY, J.N. & E. GAZTAÑAGA. 1993. Biasing and hierarchical statistics in large-scale structure. Astrophys. J. **413:** 447–452.
14. FRY, J.N., A.L. MELOTT & S.F. SHANDARIN. 1993. The three-point correlation function in an ensemble of three-dimensional simulations. Astrophys. J. **412:** 504–512.
15. FRY, J.N. & P.J.E. PEEBLES. 1978. Statistical analysis of catalogs of extragalactic objects. IX. The four-point galaxy correlation function. Astrophys. J. **221:** 19–33.
16. FRY, J.N. & R.J. SCHERRER. 1994. Skewness in large-scale structure and non-Gaussian initial conditions. Astrophys. J. **429:** 36–42.
17. FRY, J.N. & M. SELDNER. 1982. Transform analysis of the high-resolution Shane-Wirtanen catalog: The power spectrum and the bispectrum. Astrophys. J. **259:** 474–481.
18. GOROFF, M. H., B. GRINSTEIN, S.-J. REY & M. WISE. 1986. Coupling of modes of cosmological mass density fluctuations. Astrophys. J. **311:** 6–14.
19. GROTH. E. J. & P. J. E. PEEBLES. 1977. Statistical analysis of catalogs of extragalactic objects. VII. Two- and three-point correlation functions for the high resolution Shane-Writanen catalog of galaxies. Astrophys. J. **217:** 385–405.
20. GUNN, J.E. & G.R. KNAPP. 1992. *In* Sky Surveys, Protostars to Protogalaxies. T. Soifer, Ed. Astron. Soc. of Pacific Conference Series No. 43: 267.
21. HAMILTON, A.J.S. 1988. Evidence for biasing in the CfA survey. Astrophys. J. Lett. **331:** L59–L62.
22. HIVON, E., F.R. BOUCHET, S. COLOMBI & R. JUSZKIEWICZ. 1995. Redshift distortions of clusteing: A Lagrangian approach. Astron. Astrophys. **298:** 643–660.

23. JAIN, B. & E. BERTSCHINGER. 1994. Second order power spectrum and nonlinear evolution at high redshift. Astron. Astrophys. **431:** 495–505.
24. JUSZKIEWICZ, R., F.R. BOUCHET & S. COLOMBI. 1993. Skewness induced by gravity. Astrophys. J. Lett. **412:** L9–L12.
25. JUSZKIEWICZ, R., D. H. WEINBERG, P. AMSTERDAMSKI, M. CHODOROWSKI & F. BOUCHET. 1995. Weakly nonlinear Gaussian fluctuations and the Edgeworth expansion. Astrophys. J. **442:** 39–56.
26. KAISER, N. 1984. On the spatial correlations of Abell clusters. Astrophys. J. Lett. **284:** L9–L12.
27. KAISER, N. 1987. Clustering in real space and in redshift space. Mon. Not. R. Astron. Soc. **227:** 1–21.
28. KRON, R.G. 1992. A wide-field telescope for high-latitude surveys — CCD multiband photometry and multifiber spectroscopy. *In* Progress in Telescope and Instrumentation Technologies, ESO Conference and Workshop Proceedings No. 42. M.-H. Ulrich, Ed.: 635–643. ESO. Garching.
29. PEEBLES, P.J.E. 1980. The Large-Scale Structure of the Universe. Princeton Univ. Press. Princeton, NJ.
30. POSTMAN, M. & M. J. GELLER. 1984. The morphology-density relation: The group connection. Astrophys. J. **281:** 95–99.
31. SAUNDERS, W., M. ROWAN-ROBINSON & A. LAWRENCE. 1992. The spatial correlation function of *IRAS* galaxies on small and intermediate scales. Mon. Not. R. Astron. Soc. **258:** 134–146.
32. SCOCCIMARRO, R. 1997. Cosmological perturbations entering the nonlinear regime. Astrophys. J. **487:** 1–17.
33. SCOCCIMARRO, R. & J. FRIEMAN. 1996. Loop corrections in nonlinear cosmological perturbations theory. II. Two-point statistics and self-similarity. Astrophys. J. **473:** 620–644.
34. SCOCCIMARRO, R., S. COLOMBI, J.N. FRY, J.A. FRIEMAN, E. HIVON & A.L. MELOTT. 1998. Non-linear evolution of the bispectrum of cosmological perturbations. Astrophys. J. **496:** in press. .
35. SHANE, C.D. & C.A. WIRTANEN. 1967. Publ. Lick Obs. 22, Part 1.
36. SMOOT, G.F. *et al.* 1992. Structure in the COBE differential microwave radiometer first-year maps. Astrophys. J. Lett. **396:** L1–L5.
37. STRAUSS, M.A., M. DAVIS, A. YAHIL & J.P. HUCHRA. 1992. A survey of *IRAS* galaxies. V. The acceleration of the local group. Astrophys. J. **385:** 421–444.
38. STRAUSS, M.A., A. YAHIL, M. DAVIS, J.P. HUCHRA & K. FISHER. 1992. A redshift survey of *IRAS* galaxies. V. The acceleration of the local group. Astrophys. J. **397:** 395–419.
39. THOMAS, D. & J. N. FRY. 1998. Astrophys. J. Submitted.

Nonlinear Cosmological Structure Formation[a]

CHUNG-PEI MA[b]

Department of Physics and Astronomy, University of Pennsylvania, Philadelphia, Pennsylvania 19104

ABSTRACT: A general discussion of the evolution and model-dependence of both the linear and nonlinear power spectrum of density fluctuations is presented. The features of the linear power spectrum in cosmological models with cold dark matter (CDM) and cold+hot dark matter (C+HDM) are reviewed, and useful analytical approximations are summarized. Cosmological numerical simulation results are then used to illustrate the process of nonlinear gravitational clustering and to compute the nonlinear power spectrum. A new analytical approximation relating the linear and nonlinear power spectrum in C + HDM models is presented.

1. INTRODUCTION

The power spectrum $P(k)$ of matter fluctuations is a fundamental quantity in cosmology. It provides the most basic statistical measure of gravitational clustering, and an accurate determination of the power spectrum is among the most important goals of every major galaxy survey. Since each cosmological model has its distinct power spectrum, one can hope to obtain crucial information such as the matter content of the Universe and the nature of the primordial fluctuations by comparing the observed power spectrum with theoretical predictions for various plausible models.

Measurements of the power spectrum require extensive galaxy surveys covering a large volume of space. It is only recently that such surveys have been completed and are starting to reveal intriguing results. For instance, much interest has been generated by the compilation of the power spectrum determined from the APM and IRAS galaxy catalogs and the effort to reconstruct the underlying linear mass power spectrum [14], [13]. The recently completed Las Campanas survey of over 23,000 galaxies has yielded a power spectrum incompatible with the standard cold dark matter (CDM) model but still consistent with a class of open CDM, cold+hot dark matter (C+HDM), and tilted CDM models [10]. Larger ongoing projects such as the CfA Century, the 2dF, and the Sloan surveys promise to provide measurements with reduced error bars and data points on larger scales for better constraints.

On the theoretical side, calculations of the power spectrum fall into two regimes: the linear and the nonlinear, which are characterized by the amplitude of the density fluctuations $\delta = \delta\rho/\rho \ll 1$ and > 1, respectively. Since the observed power spectrum spans both linear and nonlinear regimes, theoretical calculations in both regimes must be performed before all data points can be fully utilized as constraints. In the

[a]The supercomputing time for this work was generously provided by the National Scalable Cluster Project at the University of Pennsylvania and the National Center for Supercomputing Applications.

[b]E-mail: cpma@strad.physics.upenn.edu.

linear regime, the computation of the power spectrum has become standard practice. It is obtained by integrating the coupled, linearized Einstein and Boltzmann equations that describe the evolution of the metric and density perturbations throughout the history of the Universe. Several groups have written numerical codes for such calculations, and a publicly-available version is described by Ma and Bertschinger [11] and placed at http://arcturus.mit.edu. This code computes, in both synchronous and conformal Newtonian gauges, the evolution of the phase-space distributions of photons, baryons, cold dark matter, and both massless and massive neutrinos. It is therefore applicable to most CDM, C+HDM, and CDM with a cosmological constant (LCDM) models. To gain deeper insight and for convenience, analytical approximations for the linear power spectra in these models have also been published. Details about the linear power spectrum are discussed in Section 2.

Gravitational clustering, however, eventually becomes a nonlinear process. The existence of galaxies and clusters is a manifestation of the nonlinear nature of gravity. Determining the nonlinear power spectrum $P_{\rm nl}$ is therefore an important task. Not surprisingly, $P_{\rm nl}$ is more difficult to obtain than the linear power spectrum. Higher-order perturbation theories help to extend the range of validity of the linear theory to the quasi-linear regime, but the fully nonlinear power spectrum can be calculated only from numerical simulations. Thus far, theoretical predictions of $P_{\rm nl}$ have been carried out for only a few models at limited epochs because it is difficult and time-consuming to perform numerical simulations with sufficient dynamic range to allow calculations of $P_{\rm nl}$ over a wide range of scales. Section 3 summarizes some recent work and presents new analytic and numerical results for nonlinear clustering in C+HDM models.

2. LINEAR POWER SPECTRUM

The power spectrum $P(k)$ of matter fluctuations is related to the density field δ in k-space by

$$\langle\delta(\vec{k})\delta^*(\vec{k}')\rangle = P(k)\delta_D(\vec{k}-\vec{k}')\,, \tag{1}$$

where δ_D is the Dirac-delta function. The Fourier transform of the power spectrum is the two-point correlation function $\xi(r) = \langle\delta(\vec{x}_1)\delta(\vec{x}_2)\rangle$, where $r \equiv |\vec{x}_1 - \vec{x}_2|$. For a Gaussian density field (as predicted in most inflationary theories), its statistical property is entirely determined by $P(k)$.

The shape of the linear power spectrum depends on the cosmological parameters assumed in a given theory of structure formation. It is governed by the linear perturbation theory of gravitational clustering, and can be computed by integrating the coupled, linearized Einstein, Boltzmann, and fluid equations for the metric and density perturbations. Some important parameters that affect the shape of the power spectrum are: (1) *Primordial spectral index n*: $P(k) \propto k^n$. An example is the Harrison-Zeldovich [7], [17] spectrum in which $n = 1$. Most versions of inflationary models also predict nearly $n = 1$ power spectra. (2) *Matter-radiation equality time:* The equality time $t_{\rm eq}$ is defined to be the epoch in the thermal history of the Universe when the energy density in matter equals that in radiation. The equality redshift

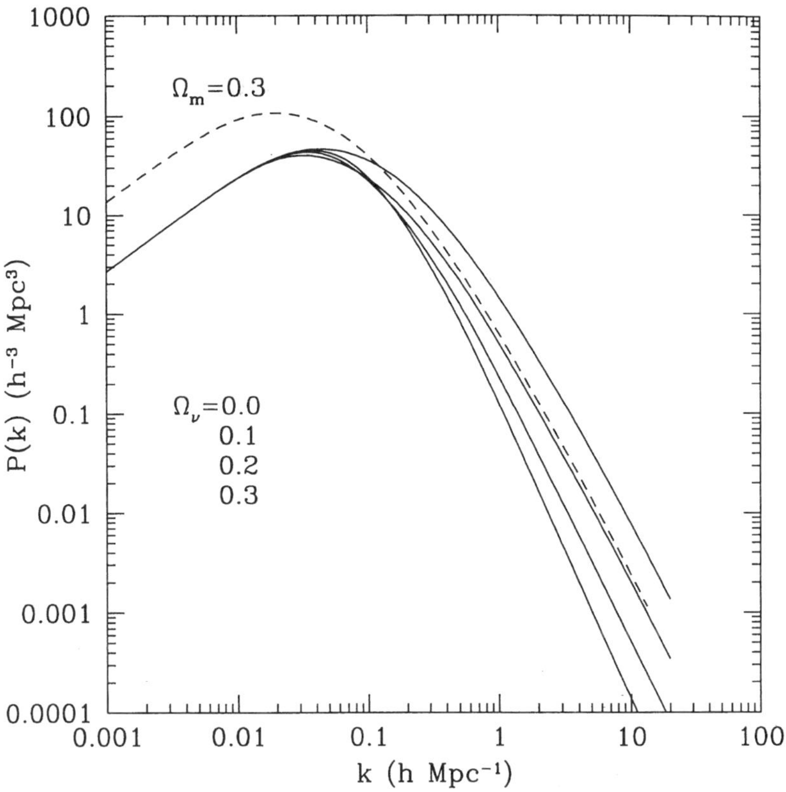

FIGURE 1. The linear power spectrum of density fluctuations at $z = 0$ for the standard CDM model (top solid), the C+HDM models with $\Omega_\nu = 0.1$, 0.2, and 0.3 (lower solid), and the LCDM model with $\Omega_m = 0.3$ and $\Omega_\Lambda = 0.7$ (dashed). All are COBE-normalized.

scales as $1 + z_{eq} \propto \Omega_m h^2$, where Ω_m is the density parameter in matter and h is the Hubble constant in units of 100 km/s/Mpc. This parameter controls the location of the peak of $P(k)$ in CDM-type models since the density perturbations with wavenumbers $k > k_{eq}$ enter the horizon in the radiation-dominated era and cannot grow appreciably, whereas perturbations with $k < k_{eq}$ enter the horizon in the matter-dominated era and can grow unimpeded. (3) *Nature of dark matter*: The pure CDM model, for example, exhibits a characteristic $P(k) \propto k^{-3}$ slope at high k, while the power spectrum for the pure hot dark matter (HDM) model is cut off exponentially below the free-streaming scale of the massive neutrinos due to the phase mixing in the neutrino phase-space distribution. The hybrid C+HDM models, which are parameterized by the neutrino mass density Ω_ν and CDM mass density $\Omega_c = 1 - \Omega_\nu - \Omega_b$ (where Ω_b is the mass density in baryons), exhibit intermediate behavior. The suppression in the power in both the cold and hot components due to neutrino free-streaming generally increases with increasing Ω_ν.

The solid curves in FIG. 1 show the linear power spectra for the pure CDM and three C+HDM models with $\Omega_\nu = 0.1$, 0.2, and 0.3. The corresponding neutrino masses in the four models are 0, 2.3, 4.6, and 7 eV, respectively. (Only one of the three species of neutrinos is assumed massive.) The models have a total matter density of $\Omega_m = 1$ and $h = 0.5$, and all are normalized to the COBE rms quadrupole $Q_{rms-PS} = 18\mu K$ [5], as evidenced by the convergence of the curves at small k. The differences at large k reflect the distinct properties of cold and hot dark matter: the larger Ω_ν is, the weaker is the gravitational clustering on small length scales. For comparison, a COBE-normalized low-density model with $\Omega_m = 0.3$, a cosmological constant $\Omega_\Lambda = 0.7$, and $h = 0.75$ is also shown (dashed curve).

Analytical approximations with less than 10% error are very convenient as an input for many calculations. The pure CDM model is well-approximated by

$$P_c(q, a, \Omega_\nu = 0) =$$

$$Aa^2k^n \left[\frac{\ln(1+\alpha_1 q)}{\alpha_1 q}\right]^2 \frac{1}{[1+\alpha_2 q+(\alpha_3 q)^2+(\alpha_4 q)^3+(\alpha_5 q)^4]^{1/2}}, \quad q = \frac{k}{\Gamma h}, \tag{2}$$

where A is the overall normalization factor, a is the expansion factor, n is the primordial spectral index, k is the wavenumber in units of Mpc^{-1}, and $\alpha_1 = 2.34$, $\alpha_2 = 3.89$, $\alpha_3 = 16.1$, $\alpha_4 = 5.46$, and $\alpha_5 = 6.71$ [1]. The normalization A can be determined from the 4-year data from the COBE satellite, and $A = 2420h^{-4}\ \mathrm{Mpc}^4$ for $n = 1$ flat models [2]. The shape parameter Γ characterizes the dependence on cosmological parameters, and a good approximation is found to be given by $\Gamma = \Omega_m h/\exp[\Omega_b(1 + 1/\Omega_m)]$ [4], [16]. The error in the approximation for the standard CDM model with $\Omega_b = 0.05$, for example, is <10%. A higher-accuracy fit for this much-studied model can be achieved by setting $\Omega_b = 0$ in Eq. (2) (i.e., setting $\Gamma = \Omega_m h$) and modifying the coefficients to $\alpha_1 = 2.205$, $\alpha_2 = 4.05$, $\alpha_3 = 18.3$, $\alpha_4 = 8.725$, and $\alpha_5 = 8.0$. The fractional error relative to the direct numerical result is then reduced to smaller than 1% for $k < 40\ h\ \mathrm{Mpc}^{-1}$ [12].

The linear power spectra for the C+HDM models require additional treatment since the effect of neutrino free-streaming on the shape of the power spectrum is both time- and scale-dependent. It is found [12] that by introducing a second-shape parameter

$$\Gamma_\nu = a^{1/2}\,\Omega_\nu h^2 \tag{3}$$

to characterize the neutrino free-streaming distance, accurate approximations to the linear power spectra in C+HDM models can be obtained. The cold and hot components clearly have different power spectra due to their different thermal properties. For the CDM component in C+HDM models, a good approximation is given by

$$P_c(k, a, \Omega_\nu) = P_c(k, a, \Omega_\nu = 0)\left(\frac{1 + b_1 x^{b_4/2} + b_2 x^{b_4}}{1 + b_3 x_0^{b_4}}\right)^{\Omega_\nu^{1.05}}, \tag{4}$$

$$x = \frac{k}{\Gamma_\nu}, \quad x_0 = x(a = 1),$$

where $P_c(k, a, \Omega_\nu = 0)$ for the pure CDM model is given by Eq. (2), and $b_1 = 0.01647$, $b_2 = 2.803 \times 10^{-5}$, $b_3 = 10.90$, and $b_4 = 3.259$ for k in units of Mpc^{-1}. The functional form for the ratio $P_c(k, a, \Omega_\nu)/P_c(k, a, 0)$ is chosen to have the asymptotic behavior $\propto a^{2(f_\infty - 1)}$which can be derived analytically, and

$$f_\infty = (5\sqrt{1 - 24\Omega_\nu/25} - 1)/4$$

is the asymptotic growth rate for $k \to \infty$, where $1 - f_\infty \propto \Omega_\nu{}^{1.05}$ is a good approximation. For the HDM component in C+HDM models, an accurate approximation is given by

$$P_\nu(k, a, \Omega_\nu) = P_c(k, a, \Omega_\nu)\left(\frac{e^{-c_1 x}}{1 + c_2 x^{1/2} + c_3 x + c_4 x^{3/2} + c_5 x^2}\right), \quad x = \frac{k}{\Gamma_\nu h}, \tag{5}$$

where $c_1 = 0.0015$, $c_2 = -0.1207$, $c_3 = 0.1015$, $c_4 = -0.01618$, and $c_5 = 0.001711$ for k in units of Mpc^{-1}.

It is often useful to have an analytic approximation for the density-weighted power spectrum $P(k) = \{\Omega_\nu P_\nu{}^{1/2} + (1 - \Omega_\nu)P_c{}^{1/2}\}^2$ that measures the total gravitational fluctuations contributed by the two components. Here the CDM and baryons have been assumed to have the same power (i.e., $P_c = P_b$), which is a good approximation for the range of redshifts and Ω_b studied in this paper. The functional form used for the CDM spectrum P_c in Eq. (4) works well here, and a good approximation for the density-weighted power spectrum in C+HDM models is given by

$$\frac{P(k, a, \Omega_\nu)}{P(k, a, \Omega_\nu{=}0)} = g(x, \Omega_\nu) = \left(\frac{1 + d_1 x^{d_4/2} + d_2 x^{d_4}}{1 + d_3 x_0^{d_4}}\right)^{\Omega_\nu^{1.05}},$$

$$x = \frac{k}{\Gamma_\nu}, \quad x_0 = x(a = 1), \tag{6}$$

where $P_c(k, a, \Omega_\nu = 0)$ again is given by Eq. (2), and the coefficients are $d_1 = 0.004321$, $d_2 = 2.217 \times 10^{-6}$, $d_3 = 11.63$, and $d_4 = 3.317$ for k in units of Mpc^{-1}.

It should be noted that the C+HDM power spectra do not obey the simple evolution $P(k) \propto a^2$ in Eq. (2) for the flat pure CDM model. The free-streaming of massive neutrinos slows down with time, allowing the neutrinos to cluster gravitationally and the neutrino density perturbations to grow on increasingly smaller scales. As a result, the growth of the C+HDM power spectra is both scale- and time-dependent. This effect is taken care of by the time-dependent parameter Γ_ν in Eq. (3), which is built in in Eqs. (4)–(6) via the scaled variable x.

3. NONLINEAR GRAVITATIONAL CLUSTERING

3.1. Numerical Simulations

The linear theory of gravitational clustering is an elegant and powerful theory. It describes accurately the growth of density perturbations from the early Universe un-

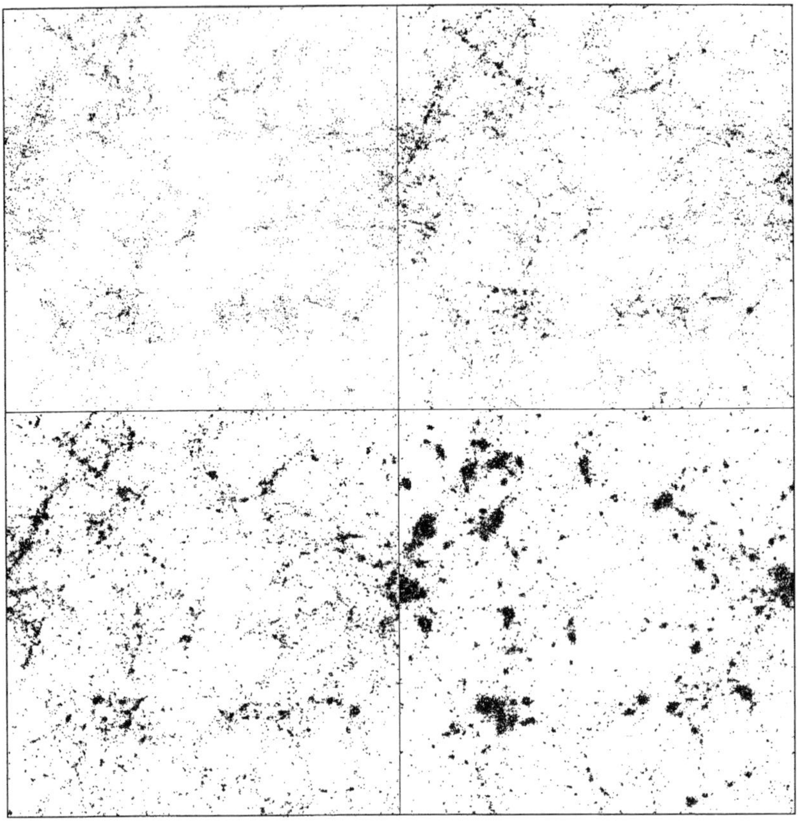

FIGURE 2. The smoothed matter distribution at four redshifts, $z = 5$ (upper left), 3 (upper right), 1.5 (lower left), and 0 (lower right), from an N-body simulation of the $\Omega_\nu = 0.1$ C + HDM model. Each panel is 100 × 100 Mpc comoving, and shows projection along one axis of the entire simulation box.

til a redshift of ~100, and the solution to the theory involves straightforward time-integration of a set (albeit a large set) of coupled ordinary differential equations. Gravitational clustering, however, eventually becomes a nonlinear process, and the study of the fully nonlinear process ultimately relies on numerical simulations.

Cosmological simulations can be roughly divided into two categories: (1) N-body simulations that deal only with dissipationless gravitational interactions of dark matter; and (2) hydrodynamical simulations that model gaseous dissipation via cooling and heating processes in addition to the gravitational interactions among the dark matter and gas. Depending on the models, the simulations are started at a redshift between 20 and 100 when the rms density fluctuations are well below unity, and the initial conditions of the simulations are generated from the linear power spectrum at that epoch.

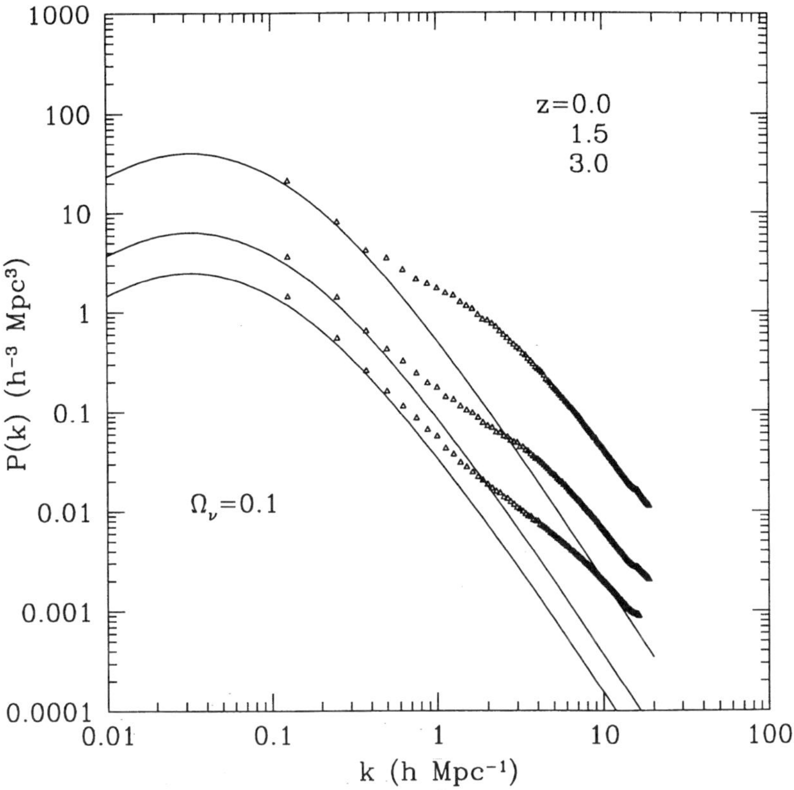

FIGURE 3. The corresponding power spectrum at z = 3, 1.5, and 0 (from bottom up), for the $\Omega_\nu = 0.1$ C+HDM model shown in FIG. 2. The solid curves show the linear $P(k)$ predicted by the linear perturbation theory; the triangles show the nonlinear $P(k)$ computed from the N-body simulation.

As an illustration of gravitational clustering in collisionless N-body simulations, FIG. 2 shows the projection of the smoothed matter distribution at four redshifts, z = 5, 3, 1.5, and 0, from an N-body simulation of the $\Omega_\nu = 0.1$ C+HDM model. The model assumes $\Omega_c = 0.85$, $\Omega_b = 0.05$, and $h = 0.5$, and is normalized to the COBE quadrupole $Q_{\rm rms-PS} = 18\mu K$ [5]. The simulation box is 100 Mpc comoving on a side, and the force softening is 50 kpc. A total of 128^3 cold and 128^3 hot particles are used. Each panel in FIG. 2 is 100× 100 Mpc comoving, and shows the projection along one axis of the entire simulation box. Periodic boundary conditions are adopted in the simulation, so for example, the dense clumps near the central left and right edges of the box in the z = 0 panel (lower right) are part of the same cluster of dark matter halos. The darkest several halos at z = 0 all have masses above $10^{14} M_\odot$.

The power spectrum of the density fluctuations offers the lowest-order measurement of the gravitational clustering exhibited in FIG. 2. In FIG. 3, we plot the corresponding nonlinear power spectra computed from the particle spatial distributions at

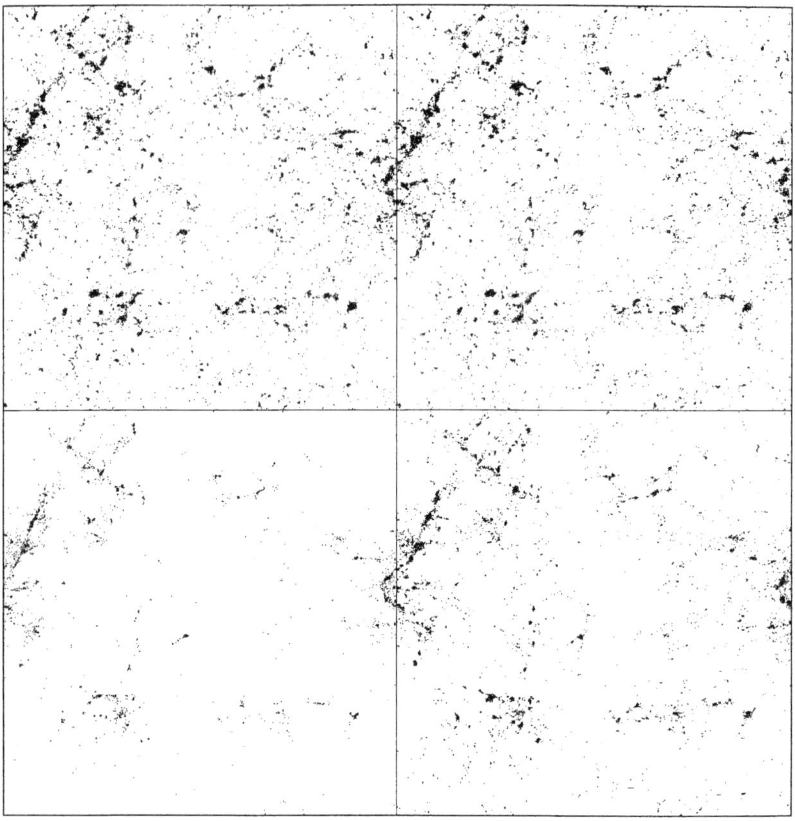

FIGURE 4. The smoothed matter distribution at z = 1.5 from N-body simulations of four cosmological models: LCDM with $\Omega_m = 0.3$ and $\Omega_\Lambda = 0.7$, and C+HDM with Ω_ν = 0.1, 0.2, and 0.3 (clockwise from upper left). Each panel is 100 × 100 Mpc comoving, and shows projection along one axis of the entire simulation box. The same initial random phases were used in all four simulations, so structures appear in similar locations in all panels.

z = 3, 1.5, and 0 shown in FIG. 2. The linear power spectra given by Eq. (6) are also shown for comparison. The hierarchical, or "bottom-up," nature of gravitational collapse in these models is evident: the high-k modes (i.e., small length scales) have become strongly nonlinear, while the low-k modes are still following the linear power spectrum. The fact that the three lowest k modes are still linear at z = 0 ensures that our choice of the simulation box size (100 Mpc) is large enough to include all waves that have gone nonlinear at the present epoch. It is also interesting to note that the point of departure from linearity moves to the left of the figure as z decreases, indicating that objects become nonlinear on increasingly larger length scales as the Universe evolves.

It is also instructive to compare the nonlinear structures at the same epoch in different cosmological models. FIGURE 4 shows the smoothed matter distribution at $z = 1.5$ from simulations of the four models in FIG. 1: a LCDM, and C+HDM with $\Omega_\nu = 0.1$, 0.2, and 0.3 (clockwise starting from upper left). The LCDM model has $\Omega_m = 0.3$, $\Omega_\Lambda = 0.7$, and $h = 0.75$, and the three C+HDM models have $h = 0.5$, $\Omega_b = 0.05$, and $\Omega_c + \Omega_\nu + \Omega_b = 1$. All models are COBE-normalized, and all simulations are performed with the same box size and force resolution as the $\Omega_\nu = 0.1$ run described earlier. (The only exception is that the $\Omega_\nu = 0.2$ and 0.3 simulations used 10×128^3 instead of 128^3 particles to represent the HDM component. We have performed tests to verify that $P(k)$ is insensitive to the number of HDM particles used in the simulations.) The initial conditions for all four simulations are generated with the same random phases; structures, therefore, appear at similar locations in all panels. The effect of Ω_ν on structure formation, however, is striking: the higher Ω_ν is, the fewer collapsed objects there are at a given epoch. This trend simply reflects the

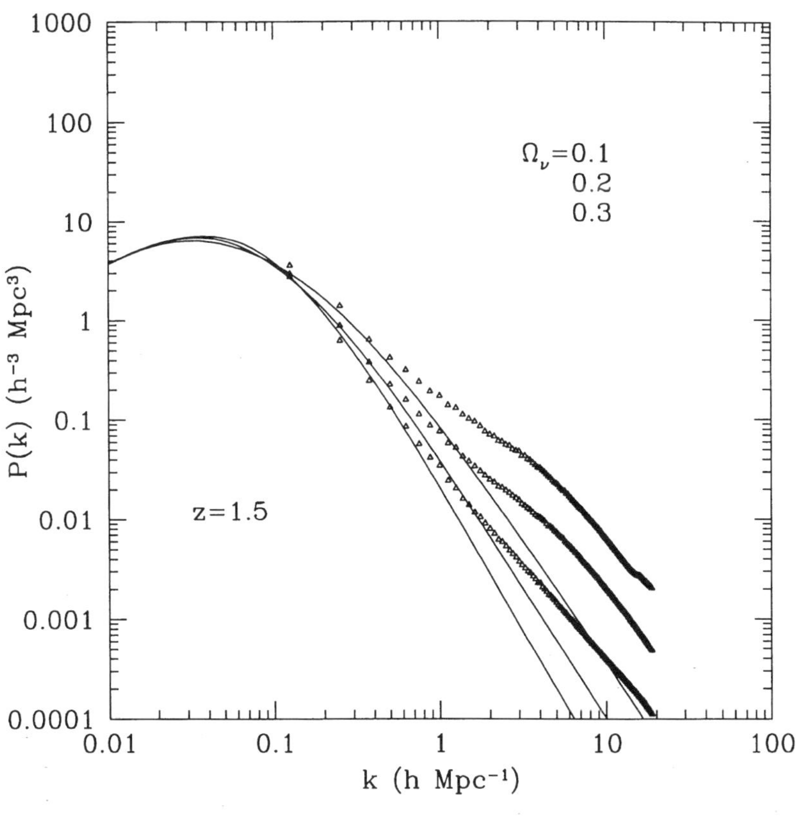

FIGURE 5. The matter power spectrum at $z = 1.5$ for the three C+HDM models shown in FIG. 4. The solid curves show the linear $P(k)$ predicted by the linear perturbation theory; the triangles show the nonlinear $P(k)$ computed from the N-body simulations.

decrease in the small-scale linear power shown in FIG. 1, which results from neutrino free-streaming. Note that the model with $\Omega_\nu = 0.3$ (lower left panel), which corresponds to a neutrino mass of 7 eV, has very few structures even at $z = 1.5$. It is also interesting to note that the upper two panels of FIG. 4 for the LCDM and $\Omega_\nu = 0.1$ C+HDM models look very similar. This is because the linear power spectra for the two models are in fact very similar at $k > 0.1\ h$ Mpc^{-1} as shown in FIG. 1. The corresponding linear and nonlinear power spectra at $z = 1.5$ for the three C+HDM models in FIG. 4 are shown in FIG. 5.

The power spectra shown thus far for the C+HDM models are the density-averaged $P(k) = \{\Omega_\nu P_\nu^{1/2} + (1 - \Omega_\nu) P_\nu^{1/2}\}^2$, where P_ν and P_c are the individual power spectra for the hot and cold components. (It has been assumed $P_c = P_{\text{baryon}}$, which is a good approximation for the redshift range of interest here.) The two components evolve distinctly due to their different thermal velocities, and the details of the shape

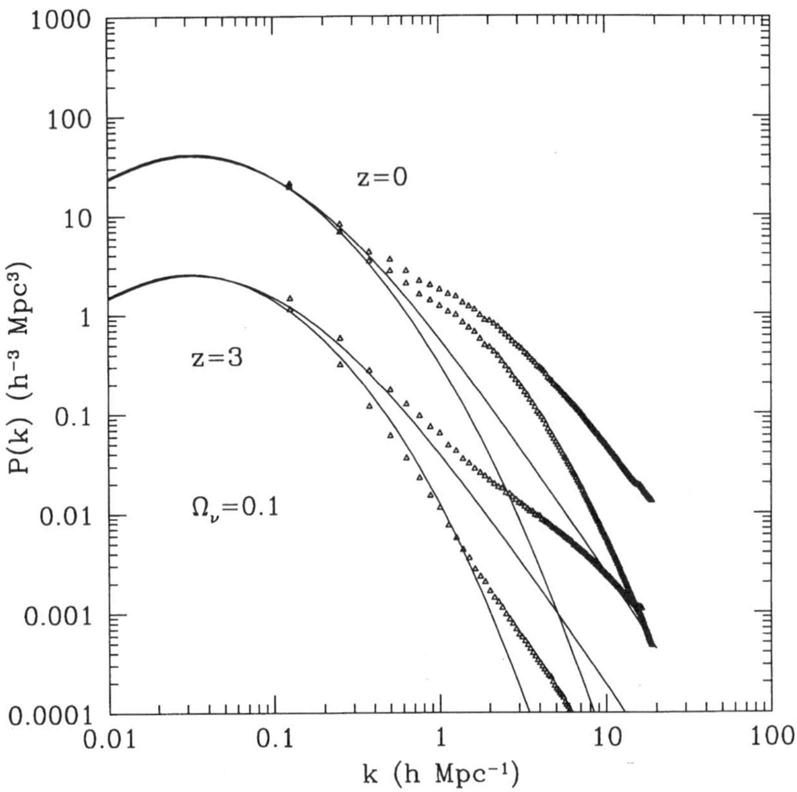

FIGURE 6. The separate CDM and HDM power spectra at z = 3 (lower set) and 0 (upper set) for the $\Omega_\nu = 0.1$ C+HDM model. The linear and nonlinear $P(k)$ are shown as solid curves and triangles, respectively. For each pair of curves, the higher one is for the CDM and the lower one for the HDM.

and the growth of the linear power spectra are discussed by Ma [12]. Here, the nonlinear power spectra for the separate components are presented in FIG. 6. For clarity, only two epochs, $z = 3$ and 0, are plotted. Notice how at $z = 3$, the HDM spectrum remains linear to almost $k = 2\ h$ Mpc^{-1} but the CDM remains linear to only $k \sim 0.5$ h Mpc^{-1}, while at $z = 0$, both components become nonlinear at $k \approx 0.3\ h$ Mpc^{-1}. This is largely due to the slowing down of the neutrinos, which makes it easier for the neutrino particles to fall into the CDM potential wells at later times, and therefore become nonlinear at smaller k. More precisely, the median thermal velocity of neutrinos is

$$v \approx 3(1+z)\frac{k_B T_{0,\,\nu}}{m_\nu c} = 15(1+z)\frac{10\text{eV}}{m_\nu}\ \text{km/s}, \tag{7}$$

where $T_{0,\nu}$ is the neutrino background temperature today, so the neutrinos in the $\Omega_\nu = 0.1$ model ($m_\nu = 2.3$ eV) have slowed down by a factor of 4 since $z = 3$ to about 60 km/s today.

3.2. Analytical Approximations

Just as it was useful to cast the linear power spectra in simple functional forms for a wide range of models (see Eqs. (2)–(6)), analytical approximations to the nonlinear power spectrum can also provide valuable insight to the process of gravitational clustering. It is, however, more difficult to compute the nonlinear P_{nl} because the numerical simulations that have sufficient dynamic range to allow calculations of P_{nl} over a wide range of scales are generally much more CPU-intensive than the integration of the linearized Boltzmann equations. The behavior of the power spectrum in the nonlinear regime therefore is less well-understood.

An early attempt to relate linear and nonlinear quantities was focused on the spatially-averaged two-point correlation function $\bar{\xi}(r)$ in models with $\Omega_{\text{m}} = 1$ and a power-law power spectrum [6]. It was found that if a given nonlinear scale r_{nl} is identified with its precollapsed linear scale r_1 by $r_1^{\,3} = r_{\text{nl}}^{\,3}\,(1 + \bar{\xi}_{\text{nl}})$, then there exists a simple, universal function relating the linear and nonlinear two-point correlation function:

$$\bar{\xi}_{\text{nl}}(r_{\text{nl}}) = F[\bar{\xi}_1(r_1)]. \tag{8}$$

At that time, this transformation appeared to be magically independent of the spectral index n of the linear power spectrum assumed in the model, and the generality of this formula rendered the task of reconstructing the primordial spectrum from the observed nonlinear clustering of galaxies less complicated. However, further tests against numerical simulations showed that this formula worked well only for spectral index $n > -1$, and it erred by factors up to 3 and 10 for $n = -1.5$ and -2, respectively [8]. In particular, the nonlinear $\bar{\xi}_{\text{nl}}$ was found to rise more sharply with increasing $\bar{\xi}_1$ for models with more negative n, so Jain *et al.* introduced an n-dependent formula to accommodate this feature. For the more realistic cosmological models such as the CDM where the spectral index is a function of scale, they proposed to use an effective index n_{eff}, defined to be $n_{\text{eff}} = d\ln P/d\ln k_0$, where k_0 is the scale at which the rms mass fluctuation $\sigma(R = 1/k_0)$ is unity. The index n_{eff} therefore rep-

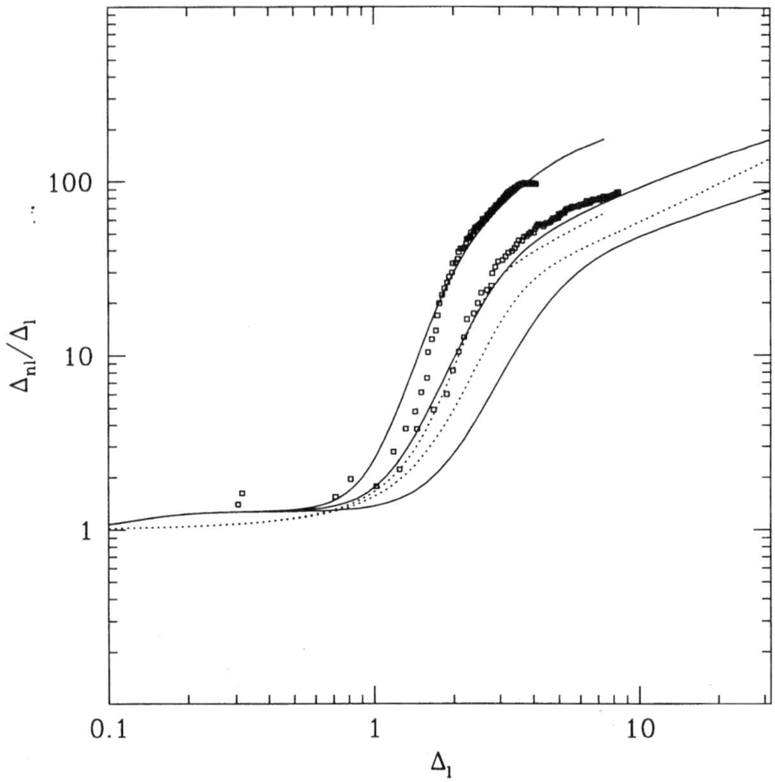

FIGURE 7. The squares show the ratio of the nonlinear and linear $\Delta(k) = 4\pi k^3 P(k)$ at $z = 0$ from the $\Omega_\nu = 0.2$ (left curve) and 0.1 (right curve) C+HDM simulations. The formula proposed by Peacock and Dodds [15] for the pure CDM models leads to large errors when applied to the C+HDM models (dotted curves; left for $\Omega_\nu = 0.2$ and right for 0.1). The solid curves show our improved analytical approximation from Eq. (10), where the rightmost one is for the pure CDM model.

resents the slope of the power spectrum at a scale where nonlinear effects are becoming important. Later work extended the analysis to models with varying matter density Ω_m and cosmological constant Ω_Λ [15], where instead of using an effective index for all scales, the local slope $n(k) = d\ln P/d\ln k$ was adopted.

Here we examine the nonlinear mapping of the power spectrum in the C+HDM models, which has not been explored yet. FIGURE 7 shows how the nonlinear density variance $\Delta_{\rm nl} = 4\pi k_{\rm nl}^3\, P_{\rm nl}(k_{\rm nl})$ diverges from the linear $\Delta_{\rm l} = 4\pi k_{\rm l}^3 P_{\rm l}(k_{\rm l})$, where the linear and nonlinear wavenumbers are related by $k_{\rm l}^3 = k_{\rm nl}^3/(1 + \Delta_{\rm nl})$. The squares are obtained from simulations of the $\Omega_\nu = 0.2$ (left curve) and 0.1 (right) C+HDM models. The curves generally obey the asymptotic condition $\Delta_{\rm nl} = \Delta_{\rm l}$ for $\Delta_{\rm l} \ll 1$, and $\Delta_{\rm nl} \propto \Delta_{\rm l}^{1.5}$ in the highly-nonlinear, stable clustering regime. The mapping, however, is clearly *not* independent of cosmology: the larger Ω_ν is in a model, the faster

$\Delta_{\rm nl}/\Delta_{\rm l}$ increases at $\Delta_{\rm l} \sim 1$. This trend reflects the n-dependence pointed out by Jain *et al.* [8] and can be explained by the different shapes of P_1 shown in FIG. 1, where the models with higher Ω_ν have less power and hence more negative slope at high k. Since all C+HDM models are normalized to COBE and have the same amplitude at low k, the critical k_0 (where $\sigma(k_0) = 1$) increases for larger Ω_ν. The effective index $n_{\rm eff}$ defined at k_0 is therefore more negative for larger Ω_ν resulting in the steeper rise of $\Delta_{\rm nl}$ in FIG. 7.

However, neither formula proposed by Jain *et al.* [8] or Peacock and Dodds [15] can be extended to the C+HDM models. The dotted curves in FIG. 7 illustrate the large discrepancies in the Peacock-Dodds fitting function when it is applied to the $\Omega_\nu = 0.1$ and 0.2 models. In retrospect, it is not surprising their formulas do not apply: Although the power spectrum at a given epoch is determined entirely from the spatial distribution of the particles, the evolution of the power spectrum depends on the particle velocities as well. A general formula for the linear to nonlinear transformation therefore must depend on both the shape and the growth rate of $P(k)$. Since both formulas are designed for models without massive neutrinos and depend only on the shape of $P(k)$, they cannot be applied to C+HDM models.

Here we propose a new analytical approximation for the mapping of linear and nonlinear power spectrum in CDM as well as C+HDM models:

$$\Delta_{\rm nl}(k_{\rm nl}) = \Delta_{\rm l}(k_{\rm l}) G[\Delta_l(k_l)], \tag{9}$$

where

$$G(x) = \frac{1 + a_1 x^4 + a_2 x^8 / g^{2.5}}{1 + a_3 x^4 + a_4 x^{7.5} / g^2}, \tag{10}$$

the coefficients are $a_1 = 4756$, $a_2 = 384.6$, $a_3 = 3732$, and $a_4 = 24.20$, and g is given by Eq. (6). The functional form of G is chosen to give the appropriate asymptotic behavior $\Delta_{\rm nl} \to \Delta_{\rm l}$ in the linear regime ($x \ll 1$) and $\Delta_{\rm nl} \propto {\Delta_{\rm l}}^{3/2}$ in the stable clustering regime ($x \gg 1$). The dependence of G on Ω_ν comes from the function g of Eq. (6), which gives the relative amplitudes of the power spectra in the C+HDM models and the pure CDM model. This function is analogous to the commonly-used growth factor $g(a = 1, \Omega_{\rm m}, \Omega_\Lambda) = \frac{5}{2}\Omega_{\rm m}[{\Omega_{\rm m}}^{4/7} - \Omega_\Lambda + (1 + \Omega_{\rm m}/2)(1 + \Omega_\Lambda/70)]^{-1}$ for LCDM models [9], [3]. They differ, however, in that the growth factor is scale-independent in LCDM models but is a function of scale in C+HDM models since neutrinos only affect the growth below the free-streaming scale, as discussed in Section 2.

4. SUMMARY

The power spectrum is the fundamental measure of gravitational clustering in cosmology. We have discussed the features and evolution of both the linear and nonlinear power spectra in various cosmological models.

The linear power spectrum P_l is calculated from time integration of the coupled, linearized Einstein, Boltzmann, and fluid equations for the metric and density perturbations. The shape and growth of the power spectrum depend on cosmological parameters such as the total matter density Ω_m, the neutrino fraction Ω_ν and the Hubble constant h. Simple analytical functions were presented that can approximate the linear P_l in both CDM and C+HDM models with less than 10% error.

The fully nonlinear power spectrum P_{nl} has to be computed from numerical simulations. Simulation output for CDM, LCDM, and C+HDM models was presented to illustrate the sensitive dependence of structure formation and evolution on cosmological parameters. We also discussed the present understanding of the relation between the linear and nonlinear power spectra, and proposed a new approximation which applies to CDM as well as C+HDM models.

REFERENCES

1. BARDEEN, J.M., JR. BOND, N. KAISER & A.S. SZALAY. 1986. Astrophys. J. **304:** 15.
2. BUNN, E. & M. WHITE. 1997. Astrophys. J. **480:** 6.
3. CARROLL, S., W. PRESS & E.L. TURNER. 1992. ARAA **30:** 499.
4. EFSTATHIOU, G., J.R. BOND, & S.D.M. WHITE. 1992. Mon. Not. Roy. Astro. Soc. **258:** 1p.
5. GORSKI, K.M. *et al.* 1996. ApJ **464:** L11.
6. HAMILTON, A.J.S., P. KUMAR, E. LU & A. MATTHEWS. 1991. Astrophys. J. **374:** Ll.
7. HARRISON, E.R. 1970. Phys. Rev. D. **1:** 2726.
8. JAIN, B., H.J. MO & S.D.M. WHITE. 1995. Mon. Not. Roy. Astro. Soc. **276:** L25.
9. LAHAV, 0., P. LIlJE, J. PRIMACK & M. REES. 1991. Mon. Not. Roy. Astro. Soc. **251:** 128.
10. LIN, H. *et al.* 1996. Astrophys. J. **471:** 617.
11. MA, C.-P. & E. BERTSCHINGER. 1995. Astrophys. J. **455:** 7.
12. MA, C.-P. 1996. Astrophys. J. **471:** 13.
13. PEACOCK, J.A. 1997. MNRAS **284:** 885.
14. PEACOCK, J.A. & S.J. DODDS. 1994. MNRAS **267:** 1020.
15. PEACOCK, J.A. & S.J. DODDS. 1996. MNRAS **280:** L19.
16. SUGIYAMA, N. 1995. Astrophys. J. S. **100:** 281.
17. ZELDOVICH, YA. B. 1972. MNRAS **160:** 1p.

On the Large-Scale Structure of Hickson Compact Groups of Galaxies

RENATO A. DUPKE AND WENTAO WU

Department of Physics and Astronomy, University of Alabama, Box 870324,Tuscaloosa, Alabama 35487-0324

ABSTRACT: We present a study on the integral two-point correlation functions of Hickson compact groups of galaxies. The results show no evidence that these objects form large-scale structures beyond $10h^{-1}$ Mpc. Their clustering is much weaker than that expected from the universal dimensionless correlation function for galaxy associations, but is overall more similar to that of galaxies. The results suggest that either Compact Groups are not real physical systems or their "seeding" is very different from that of other galaxy associations.

1. INTRODUCTION

Compact groups of galaxies (CG) are usually characterized by a very high projected galaxy density (4 or more galaxies in a typical projected separation of $100h^{-1}$kpc, low velocity dispersions (on the order of the stellar velocity dispersion within a galaxy). These characteristics indicate a short collapse time for the system. If the galaxies in a compact group are gravitationally bound they should represent a transient phase between loose associations and giant elliptical galaxies, which would appear as consequence of galaxy merging [1], [2]. The fact that the inferred frequency of merging in compact groups is smaller than that predicted by models [3], [4] together with the high incidence of redshift discordant members in CGs, suggests another possibility, that is, CGs are merely the effect of galaxy projections [5]–[8].

A fundamental step to understand the nature of CGs is to know the large-scale structures formed by CGs themselves. The two-point correlation function (TPCF) is the minimum constraint that any cosmological model predicting the formation of extragalactic structures should satisfy. In calculating the spatial correlation functions two methods were used in order to take into account sample boundary problems. We also adopt the standard cosmological model with $\Omega = 1$, $\Lambda = 0$ (flat universe) and we neglect the gravitational evolution, in comoving space, of large-scale structures on the TPCFs. Spatial positions, distances and separations are described in comoving space.

2. THE STATISTICAL SAMPLE OF COMPACT GROUPS

Our sample consists of 96 groups of galaxies taken from the Hickson compact group (HCG) catalogue [9]–[11]. HCGs were selected from the E-plates of the Palomar Observatory Sky Survey satisfying the following selection criteria: $N_{CG} \geq 4$ galaxies (group richness criterion) $\Delta m \leq 3$ magnitudes (magnitude concordance criterion) $D_g/R_{CG} \geq 3$ (isolation criterion) $\langle\mu_{CG}\rangle \leq 26.0$ mag arcsec^{-2} (compactness criterion) where N_{CG} is the number of galaxy members in a group; Δm is the difference of magnitudes between the faintest and brightest galaxy members; D_g is the distance from the centroid of the group to the nearest nongroup galaxy; R_{CG} is the smallest radius which includes all group members. $\langle\mu_{CG}\rangle$ is the mean surface brightness within R_{CG}. We adopt the sample boundaries as $\delta(1950) \geq -33°$ (Palomar Sky Survey limit) and $|b^{II}| \geq 23°$ (Galactic disk incompleteness problem).

Due to the redshift discordant problem in HCGs, we apply certain extra exclusion criteria. We consider any two galaxies in an HCG to be redshift discordant if their difference in recession velocities is not less than 1000 km/s. Most HCGs do not have redshift discordant galaxies, and they are considered as well-defined groups. We separate the HCGs, which have redshift discordant galaxies into the following cases. In an HCG that has redshift discordant members, every complete set of redshift accordant galaxies with total number of 2 or less is excluded from our sample; every complete set of redshift accordant galaxies with total number of three or more is considered as a separate group. We call a finally selected group with only three redshift accordant galaxies a "triple system." This distinction is important because according to Hickson's group richness criterion (≥ 4 members), they would not be considered as "real" CGs, maybe representing a different population of galaxy systems.

The distance to each selected group is calculated from the average redshift of its galaxy members, using the correction to the barycenter of the *local group* as given by Yahil *et al.* [13] and corrected for the Virgocentric infall following the linear model of Schechter [14]. We took into account the galactic latitute selection effect and the red-shift selection effect. To take into account the red-shift selection effect we fit the cumulative number distribution of CGs versus CGs recession velocity (v) in our sample as

$$N(\leq v) = \frac{\alpha}{\left(1 + \left(\frac{v_0}{v}\right)^2\right)^\eta}, \tag{1}$$

where v_0, α, η are the fitting parameters.

3. METHOD

The joint probability of finding sample elements (in our case CGs) centered within two volume elements dV_1 and dV_2 at separation r is proportional to the sizes of each of the elements [15]

$$dP = n^2[1 + \xi(r)]dV_1 dV_2, \tag{2}$$

where n is the number density and ξ is the spatial TPCF which generally can be well described by the power law

$$\xi(r) = \left(\frac{r}{r_0}\right)^{-\gamma}, \tag{3}$$

where r_0 is a correlation scale length. Analogously, one can define the joint probability of finding sample elements (in our case a CG) centered within two solid angle elements $d\Omega_1$ and $d\Omega_2$ at angular separation θ in the sky as

$$dP = N^2[1 + W(\theta)]d\Omega_1 d\Omega_2, \tag{4}$$

where N is the surface number density, $W(\theta)$ is the angular TPCF generally described by

$$W(\theta) = \left(\frac{\theta}{\theta_0}\right)^{-\beta} \tag{5}$$

where $\beta = \gamma - 1$ according to Limber's equation [15]. The angular and spatial integral TPCFs are defined. respectively, by

$$\begin{aligned} \Gamma_{\text{ang}}(\theta) &= \frac{\int dP}{\int N^2 d\Omega_1 d\Omega_2} = \frac{\int_0^\theta (1 + \Omega(\theta'))d(1 - \cos\theta')}{1 - \cos\theta}, \\ \Gamma_{\text{sp}}(r) &= \frac{\int dP}{\int n^2 dV_1 dV_2} = \frac{\int_0^r (1 + \xi(r'))dV(r')}{V(r)}, \end{aligned} \tag{6}$$

where dP is given by Eqs. (3) and (5), respectively, for the spatial and angular cases, r is the comoving spatial separation and $V(r)$ is the comoving spherical volume with radius r. Note that in Eqs. (6), the integrals over volumes and solid angles are taken over regions large enough to be considered to be statistically fair representations of the large-scale structures of CGs. The integrals over dP are taken within the same regions but with the further constraint that the separations between volume and solid angle elements are $\leq r$ and θ, respectively. The integral TPCF gives the total number of pairs (any two CGs) with angular or spatial separation $\leq\theta$ or r relative to that in a homogeneous distribution. or equivalently, the average number of CGs associated with a CG, within angular or spatial separation $\leq\theta$ or r, compared to the one in a homogeneous distribution with the same average CG number density.

Using in the above definitions (Eqs. (6)) for the angular/spatial TPCFs Eqs. (3) and (5), and further expanding the trigonometric functions, one can easily see that the integral TPCFs can be expressed as follows:

$$\begin{aligned} \Gamma_{\text{sp}}(r) &= 1 + \frac{3}{3-\gamma}\left(\frac{r_0}{r}\right)^{\gamma}, \\ \Gamma_{\text{ang}}(\theta) &= \frac{1 - \cos\theta + \sum_{n=0}^{\infty} \frac{(-1)^n}{(2n+1)!}\theta_0^{\beta}\frac{\theta^{2n+2-\beta}}{2n+2-\beta}}{1 - \cos\theta}. \end{aligned} \tag{7}$$

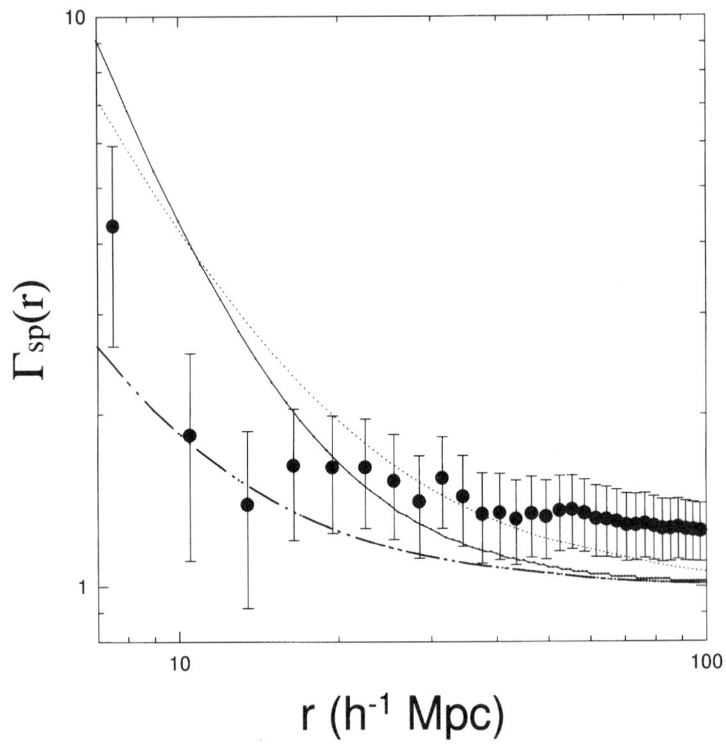

FIGURE 1a. Spatial TPCF (all CGs) Method 1.

Two methods were used to calculate the integral TPCFs: the standard method [16] and an alternative method, which was proposed to minimize the sample boundary effect. In our study, both methods were used to evaluate the average number of CGs associated with a CG, as required by the definition of the integral TPCF. In the standard method this average is taken over the whole sample. In the other method the average is taken over only small reference subsamples. The standard method assumes that the clustering properties inside and outside the sample boundaries are the same. It works well for small angular scales, but for scales comparable to the size of the sample, the hypothesis of homogeneity may be compromised. The suggested alternative method to avoid this boundary problem consists of averaging the correlations for CGs in subsamples small enough, so that one does not need to make any assumption about the distributions outside the sample itself. This method is called the reference subsample method (the application of this method to galaxy TPCFs was given by Wiedermann and Atmanspacher [17]). These two methods are discussed more extensively in Wu and Dupke [18]. Notice that the errors shown here are the random errors.

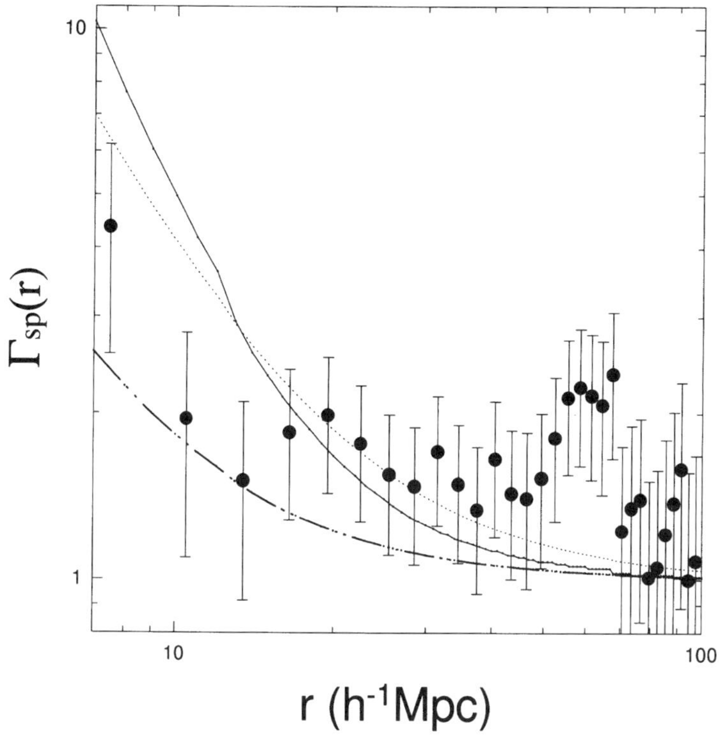

FIGURE 1b. Spatial TPCF (all CGs) method 2.

4. RESULTS AND DISCUSSION

The correlation functions were calculated for two different samples. One that includes all CGs selected by the criteria described in the sample section. The second one excludes the triple systems so that we can test if they follow the same statistical behavior of the other ("real") CGs. The spatial and angular integral TPCFs for the sample which includes the triple systems are shown in FIGS. la, 1b, 3a, and 3b for both methods. The plots for the spatial integral TPCF for the sample which excludes the triple systems are shown in FIGS. 2a and 2b.

For the cases where the triple systems are excluded the correlations are weak and there is no evidence of structures larger than 1–2°, for the angular TPCF, and $10h^{-1}$ Mpc. for the spatial TPCF. Actually, if we exclude the triple systems, the correlation is very weak and there is no evidence that compact groups form very large scale structures, angular or spatial, at all. With the inclusion of triple systems the statistical properties of compact groups change. It can be seen (FIGS. 3a and 3b) that for the sample which includes the triple systems, one can find the largest angular

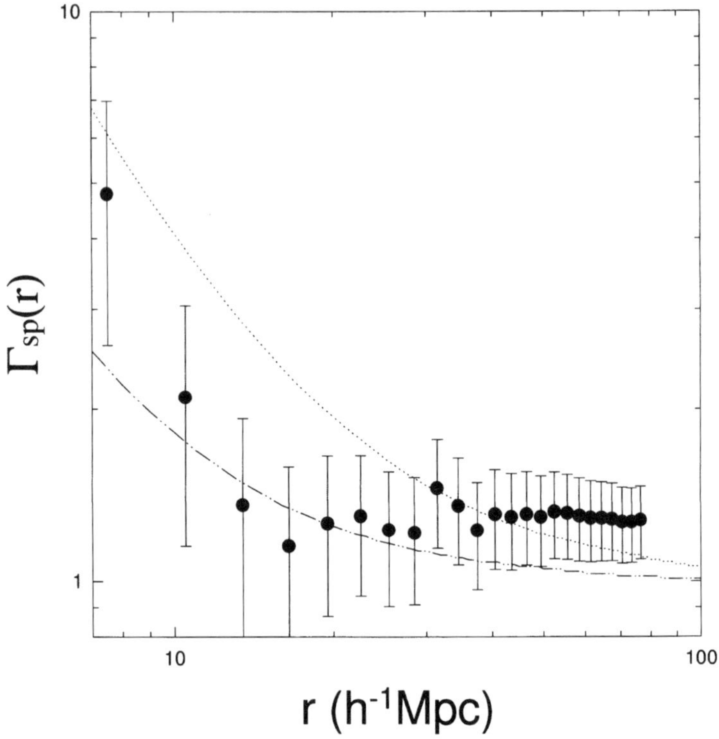

FIGURE 2a. Spatial TPCF ("real" CGs) method 1.

structures ($\Gamma \geq 1$) on the order of 20°. Besides, the angular integral TPCF for this sample is the only one well behaved among all cases. There is no evidence of spatial structures larger than approximately $10h^{-1}$ Mpc and the spatial correlation is very weak, but one can see a well-behaved (strong) angular correlation and very large angular structures, a kind of "angular conspiracy."

Assuming the angular TPCF as a power law described in Eq. (5), one can recover the parameters of the angular TPCF by fitting Eq. (7) to the angular integral TPCF. We only fitted the cases for which the triple systems were included because only they exhibit well-behaved TPCFs. In the fitting it is sufficient to consider only the first three terms of the summation in Eq. (7). The fitting results are plotted as the solid curves in FIGS. 3a and 3b. It is possible to obtain the expected spatial TPCF for the case including the triple systems from the fittings of the angular TPCF if we assume Limber's equation is valid for this case (Appendix A). Limber's equation shows that the power law form of the angular TPCF (Eq. (5)) also indicates a power law form for the spatial TPCF (Eq. (3)). The subsequently expected spatial integral TPCFs (Eq. (7)) are plotted in FIGS. 1a and 1b (solid curves). One can see that these curves are systematically higher than the actual data points within the range where the correlation is expected to be significant.

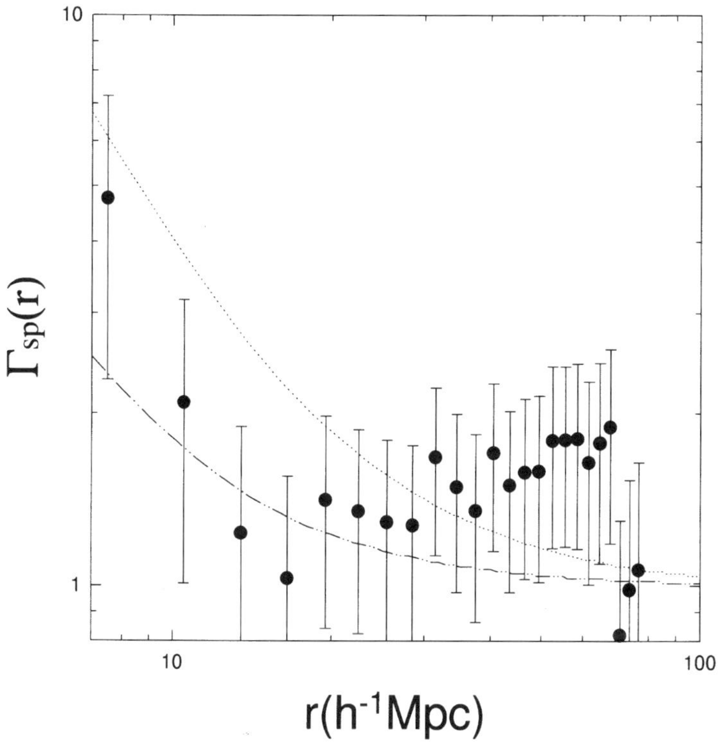

FIGURE 2b. Spatial TPCF ("real" CGs) method 2.

It is known that different populations of galaxies and galaxy associations follow a universal dimensionless correlation function (UCF) [19],

$$\xi_i = \left(\frac{0.4d_i}{r}\right)^{1.8}, \tag{8}$$

where d_i is the mean separation for a population i. For comparison, in FIGS. la, 1b, 2a, and 2b, the dashed-dotted lines and the dotted lines show, respectively, the spatial integral TPCF for galaxies, and that of CGs expected from the UCF. Both curves (galaxies and UCF) are calculated with $\gamma = 1.8$. The scale length of galaxies has been measured to be $r_0 = 4h^{-1}$ Mpc for galaxies [15]. For CGs, the UCF predicts a value of $r_0 = 11.3h^{-1}$ Mpc, which is estimated according to $r_0 = 4n^{-1/3}$ (Eq.(8)), with a mean density $n = 4 \times 10^{-5}\, h^3\, Mpc^{-3}$ [20].

It should be noted that this represents a lower limit for the integral UCF since that is the highest value for n estimated. Barnes [2] estimates $n = (0.8 - 2.4) \times 10^{-5}\, h^3\, Mpc^{-3}$ from the APM survey and Rose's catalogue, which would further increase the value of r_0 and, therefore, the value of the integral TPCFs predicted by the UCF, and

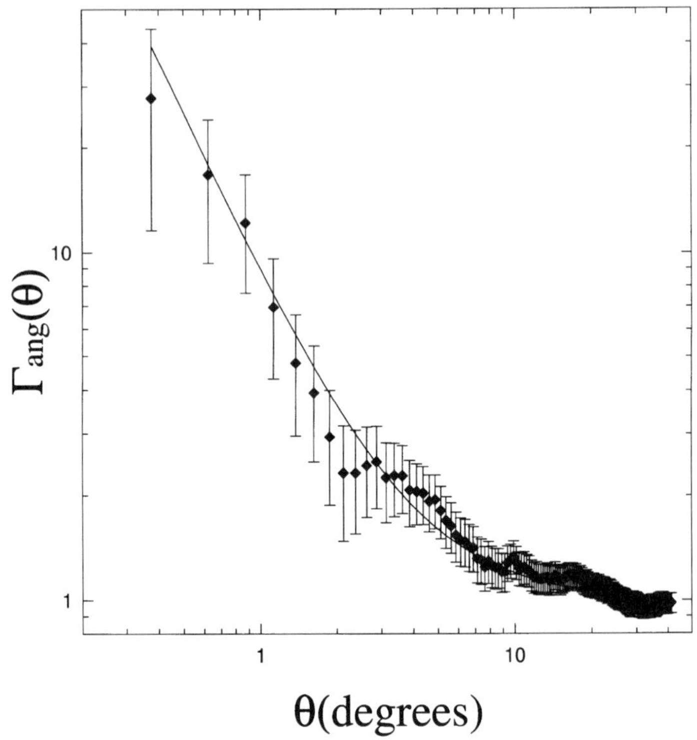

FIGURE 3a. Angular TPCF (all CGs) method 1.

thus enhancing the discrepancy between the calculated and predicted integral TPCFs. The spatial TPCF expected from the observed angular TPCF according to the Limber's equation is systematically larger than that observed at typical spatial correlation scales ($r < 15h^{-1}$ Mpc). For the case where the triple systems are excluded, the UCF predicts that the typical scales of the structures ($\Gamma(r) \approx 2 - 3$) should not be less than $10–20h^{-1}$ Mpc. However, one can see from FIGS. 2a and 2b, that at such scale, the distribution of CGs in the sample is already not different from a homogeneous one ($\Gamma(r) \approx 1$).

A general result of our paper is that the correlation of CGs is much weaker than what would be expected from the dimensionless UCF. The CG's large-scale structures are more similar to those of galaxies than to those of other galaxy associations. The angular conspiracy, i.e., objects angularly correlated but not spatially correlated, means that the Limber's equation is broken down if we include the triple systems. In using Limber's equation we have assumed that the clustering properties are spatially isotropic in the sample. The obvious explanation for the invalidity of Limber's equation, or, "angular conspiracy," is that there are large-scale nonisotropic structures in the sample. Therefore, a reasonable explanation for the "angular conspiracy," is that the triple systems, possibly together with some other groups of gal-

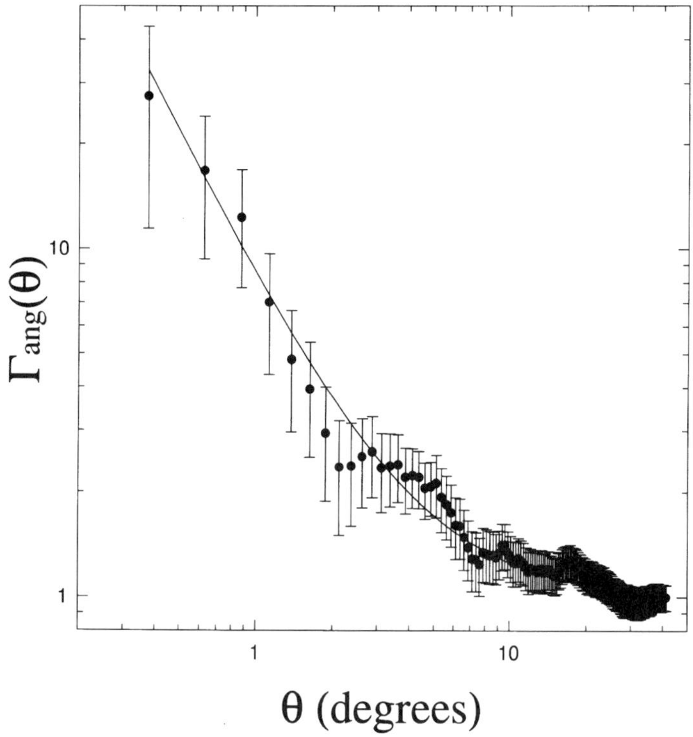

FIGURE 3B. Angular TPCF (all CGs) method 2.

axies, form filamentary structures with sizes comparable to the sample extension. Filamentary structures oriented along the line of sight would create strong angular clustering but with weak spatial clustering. In such filamentary structures the probability of galaxy projection along the filament is higher and some triple (or even double) systems would be misclassified as CGs in the HCG catalogue. This idea is in agreement with a filamentary seeding process as suggested by Hernquist *et al.* [7]. It should be noticed that throughout this work we considered that the HCG catalogue is a statistically good sample of CGs.

From our analysis we cannot say anything about the large-scale structures of triple systems alone, but it seems a reasonable hypothesis that, at least a certain number of galaxies (groups) could be formed in a filamentary-like distribution. In this case, HCGs would be alignments of pairs and triplets of galaxies which are disposed in filament-like structures. This would also be consistent with the interaction properties of galaxies observed in some compact groups (since galaxy duplets and triplets could be interacting), and with the high incidence of red-shift discordant galaxies in CGs. This idea seems to be the most promising for it is the most economic explanation regarding the problematic nature of HCGs.

APPENDIX: THE LIMBER EQUATION

In order to determine the scale length we use the relativistic Limber's equation which for a flat universe can be written as

$$W(\theta) = I_\gamma \theta^{1-\gamma}\left(\frac{r_0}{D_*}\right)^{\gamma} \frac{\int_0^\infty dX(1+z)^\gamma X^{5-\gamma}(n_0\Phi_0(X))^2}{\left[\int_0^\infty X^2 dX n_0 \Phi(X)\right]^2}, \qquad \text{(A.1)}$$

where $X = D/D_*$, and $I_\gamma = \Gamma(1/2)\Gamma(\gamma - 1/2)/\Gamma(\gamma/2)$, $\gamma = \beta + 1$, D_* is a distance scale length, D is the comoving distance, z is the red-shift, Φ_0 is the selection function, n_ϕ is the number density (mean) and Γ is the Euler gamma function. This equation assumes that there is no dynamical evolution so that the parameter ϵ in the original equation [21, Eq. (58)] is set to be equal to −3. We denote the spatial number density in the sample by Φ so that $\Phi = n_0\Phi_0$. The comoving distance is given by

$$D = \frac{2c}{H_0}\left(1 - \frac{1}{\sqrt{1+z}}\right),$$

and we set $D_* = 2c/H_0$ so that the variable of integration will be expressed as a function of red-shift by

$$X = \left(1 - \frac{1}{\sqrt{1+z}}\right), \qquad \text{(A.2)}$$

Defining the integral of the selection function over X by Φ_1,

$$\Phi_1(X) = \Omega_s K \int_0^X X^2 \Phi(X) dX, \qquad \text{(A.3)}$$

where Ω_s and K are the total solid angle covered and a normalization factor for the galactic latitute selection function, respectively, so this integral shows the total number of CGs with normalized distances smaller or equal than X, or equivalently, with red shifts smaller than z. From Eq. (A.3) one sees that

$$[X^2\Phi(X)]^2 \propto \left(\frac{d\Phi_1(X)}{dX}\right)^2. \qquad \text{(A.4)}$$

Considering that $\Phi_1(X(z))$ can be given by Eq. (1) and applying this equation together with (A.2) and (A.4) in (A.1) one gets

$$W(\theta) = C_\gamma \theta^{1-\gamma}\left(\frac{r_0}{(z_0)^{\frac{2}{\gamma}} D_*}\right)^{\gamma} = \left(\frac{\theta}{\theta_0}\right)^{-\beta}, \qquad \text{(A.5)}$$

where C_γ is given by

$$C_\gamma = 8\eta^2 I_\gamma \int_0^\infty (1+z)^{\beta+2.5}\left(1 - \frac{1}{\sqrt{1+z}}\right)^{-\beta} \frac{\left(\frac{z}{z_0}\right)^{4\eta-2}}{\left(1+\left(\frac{z}{z_0}\right)^2\right)^{2\eta+2}} dz .$$

This integral can be calculated numerically and, subsequently, so can the spatial scale length, which is given by

$$r_0 = D_* \left(\frac{C_\gamma}{\theta_0^\beta z_0^2}\right)^{-\frac{1}{1+\beta}} .$$

ACKNOWLEDGMENTS

The authors would like to thank Carlos Rabaça and Henry Kandrup for helpful discussions, Cong Xu for providing his program to correct Virgo infall motion in galaxy redshifts.

REFERENCES

1. Barnes, J.E. 1985. The dynamical state of groups of galaxies. Mon. Not. Roy. Astron. Soc. **215:** 517–536.
2. Barnes, J.E. 1989. Evolution of compact groups and the formation of elliptical galaxies. Nature. **338:** 123–126.
3. Sulentic, J.W. 1987. Properties of dense galaxy groups and the implications of their existence. Astrophys. J. **322:** 605–617.
4. Zepf, S.E. 1993. The frequency of mergers in compact groups. Astrophys. J. **407:** 448–455.
5. Mamom, G.A. 1986. Are compact groups of galaxies physically dense? Astrophys. J. **307:** 426–430.
6. Mamon, G.A. 1987. The dynamics of small group of galaxies. I. Virialized groups. **421:** 622–644.
7. Hernquist, L., N. Katz & D.H. Weinberg. 1995. Physically detached 'compact groups.' Astrophys. J. **442:** 57–60.
8. Ostriker, J. P. Lubin, L. M. & Hernquist. L. 1995. Using X-rays to determine which compact groups are illusory. Astrophys. J. **444:** L61–L64.
9. Hickson, P. 1982. Systematic properties of compact groups of galaxies. Astrophys. J. **255:** 382–391.
10. Hickson, P. 1993. Atlas of compact groups of galaxies. Astrophys. Lett. **29:** 1–3.
11. Hickson, P., E. Kindl & J.R. Auman. 1989. A photometric catalog of compact galaxies. Astrophys. J. **70:** 687–698.
12. Hickson, P., C. Mendes de Oliveira, J. Huchra & G. Palumbo. 1992. Dynamical properties of compact groups of galaxies. Astrophys. J. **399:** 353–367.
13. Yahil, A., G.A. Tammann & A. Sandage. 1977. The local group — The Solar motion relative to its centroid. Astrophys. J. **217:** 903–915.
14. Schechter, P.L. 1980. Mass-to-light ratios for elliptical galaxies. AJ. **85:** 801–811.

15. PEEBLES, P.J.E. 1980. The Large Scale Structure of the Universe. Princeton University Press.
16. PEEBLES, P.J.E. 1993. Principles of Physical Cosmology. Princeton University Press. Princeton, NJ.
17. WIEDENMANN, G.O. & H. Atmanspacher. 1996. Integral pair correlations in the angular distribution of galaxies. Astron. & Astrophys. **229:** 283–291.
18. WU. W. & R.A. DUPKE. 1998. In preparation.
19. BAHCALL, N. A. & M.J. WEST. 1992. The cluster correlation function — Consistent results from an automated survey. Astrophys. J. **392:** 419–423.
20. MENDES DE OLIVEIRA, C. & P. HICKSON. 1991. The luminosity function of compact groups of galaxies. Astrophys. J. **380:** 30–38.
21. GROTH. E.J. & P.J.E. PEEBLES. 1977. Statistical analysis of catalogs of extragalactic objects. VII. Two- and three-point correlation functions for the high-resolution Shane-Wirtanen catalog of galaxies. Astrophys. J. **217:** 385–405.

Magnetohydrodynamic Origin of Jets from Accretion Disks[a]

R.V.E. LOVELACE[b] AND M.M. ROMANOVA[c]

[b]*Department of Astronomy, Cornell University, Ithaca, NY 14853*

[c]*Space Research Institute, Profsoyuznaya 84/32, Moscow 117810, Russia; and Department of Astronomy, Cornell University, Ithaca, NY 14853*

ABSTRACT: A review is made of recent magnetohydrodynamic simulations of outflows from accretion disks for different distributions of magnetic field threading the disk. The problem of the opening of a closed field line configuration linking a magnetized star and an accretion disk is also discussed.

INTRODUCTION

A review is made of magnetohydrodynamic (MHD) theory and simulations of outflows from accretion disks for different distributions of magnetic field threading the disk. In one limit of a relatively weak, initially diverging magnetic field, both thermal and magnetic pressure gradients act to drive matter to an outflow, while a toroidal magnetic field develops, which strongly collimates the outflow. The collimation greatly reduces the field divergence and the mass outflow rate decreases after an initial peak. In a second limit of a strong magnetic field, the initial field configuration was taken with the field strength on the disk decreasing outwards to small values so that collimation was reduced. As a result, a family of stationary solutions was discovered where matter is driven mainly by the strong magnetic pressure gradient force. The collimation in this case depends on the pressure of an external medium. These flows are qualitatively similar to the analytic solutions for magnetically driven outflows [1]. The problem of the opening of a closed field line configuration linking a magnetized star and an accretion disk is also discussed.

Many compact astrophysical objects emit powerful, highly-collimated, oppositely directed jets. Included are the radio jets of active galaxies, quasars [2], and old compact stars in binaries [3], and emission line jets in young stellar objects [4], [5]. An example of the latter is shown in FIG. 1.

A broad spectrum of ideas and models have been put forward to explain astrophysical jets (see reviews by Begelman, Blandford, and Rees [6], Bisnovatyi-Kogan [7]). The matter is thought to go to the jet from the inner region of an accretion disk surrounding the compact object — a star or black hole. The disk matter must then be accelerated to a velocity higher that the escape velocity from the central object. Further, the jet matter should have sufficient momentum to propagate through surrounding inter-stellar or intra-galactic matter out to large distances. However, the main

[a]This work was supported in part by NSF grant AST-9320068, NASA grant NAGW 2293, and grant No. RP-173 from the U.S. Civialian R&D Foundation for the Independent States of the Former Soviet Union.

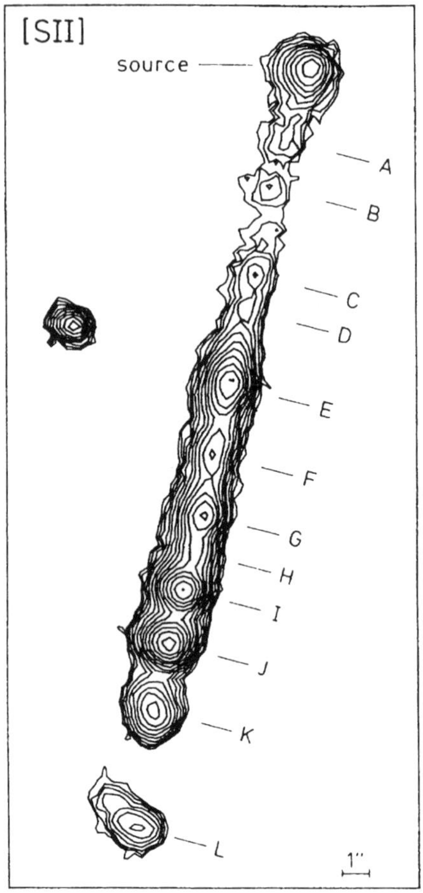

FIGURE 1. SII line intensity contour plot for the protostellar jet in the object HH 34 adapted from Figure 2 of Bührke, Mundt, and Ray [5]. The letters denote different regions of enhanced emission.

physics questions regarding the jets are *not* answered: (1) What is the main driving force pushing matter into the jet? (2) What determines the mass flow rate in the jet $\dot{M}_j$, and what fraction is this of the accretion rate? (3) What physics determines the asymptotic speed or Lorentz factor of the bulk flow? (4) What determines the collimation of the jet and at what distance from the central object does the jet become collimated? (5) What is the acceleration mechanism of leptons to Lorentz factors $\sim 10^2$ - 10^3 in the radio jets?

Here, we focus on jet formation mechanisms in which an ordered magnetic field has an essential role. Electrodynamic models for the origin of narrow jets were developed as an outgrowth of the Goldreich-Julian [8] aligned pulsar model [9]–[13].

In these models the rotation of a Keplerian accretion disk twists a poloidal field threading the disk, and this results in outflows out of the disk which carry angular momentum (in the twist of the field) and energy (in the Poynting flux) away from the disk, thereby facilitating the accretion of matter. The basic mechanism is related to that involved in the braking of stellar rotation by magnetic stellar winds [14], [15].

Recent theoretical work on the formation of jets has focused on magnetohydrodynamic (MHD) models of jet formation from accretion disks and stars. A strong case for magnetic or MHD driving of the jets of protostellar systems emerges since the temperature of the inner regions of these systems is insufficient to permit driving by thermal or radiation pressure (see, for example, Königl and Ruden [16]). Most of these investigations have been analytical or semi-analytical, and part of this has been an outgrowth from the self-similar model of Blandford and Payne [17] (see, for example, [18]–[21]). The outflows in this model are often referred to as "centrifugally driven" owing to the driving force close to the disk: If the poloidal magnetic field lines diverge from the disk surface (making an angle with the z-axis of more than 30°), then the sum of the gravitational and centrifugal forces is in the $+z$ direction for an MHD fluid particle which initially tends to maintain a constant angular rotation rate. The self-similar models are unsatisfactory in the respect that they must be cutoff at small cylindrical radii, $r \leq r_{min}(z)$. This is the most important region of the jet flow. Observations of optical stellar jets [4] reveal jet velocities ~200 – 400 km/s, which are comparable to the Keplerian disk velocity close to the star's surface. This suggests that the jets originate from the inner region of the disk close to the star (see, for example, [22], [23]).

A general treatment of stationary axisymmetric MHD flows leads to the generalized Grad-Shafranov equation which is a second-order, nonlinear partial differential equation for the flux function $\Psi(r, z) = rA_{\phi}(r, z)$, where A_{ϕ} is the toroidal component of the vector potential [24]–[30]. This equation represents an extension of the work of Grad and Rubin [31] and Shafranov [32] on plasma equilibria to include the plasma flow using the theory of Chandrasekhar [33], Woltjer [34], and Mestel [35].

An approximate method for deriving stationary jet flow solutions of the Grad-Shafranov equation has been independently developed by Koupelis and Van Horn [36] and Lovelace *et al.* [1], [37], [38]. The method involves averaging the different physical variables over the z = constant cross-section of the jet, and it gives rise to a set of ordinary nonlinear differential equations referred to as "envelope equations." These equations give the jet's axial flow velocity, radius, rotation rate, and poloidal and toroidal field components as a function of axial distance from the disk. This method captures the main physics of the Grad-Shafranov equation, the slow and fast magnetosonic points and the Alfvén singularity. The envelope equations can be evaluated using "shooting methods" to pass through the critical points. The slow critical point is found to be located close to the disk surface [1]. At this point the magnetic force $-\partial(B_{\phi}^2/8\pi)/\partial z$ is crucial in driving the flow outward and for this reason the flows were referred to as "magnetically driven" [1]. The gradient of B_{ϕ}^2 is related to the divergence of the poloidal field lines so that this driving mechanism is related to that of the above-mentioned centrifugal winds.

The analytical work on jet formation involves major simplifications and approximations. This motivates studies of the formation of centrifugally and/or magnetically driven jets by numerical 2-D and 3-D MHD simulations.

Simulation studies of jet flows have focused on three main regions: (1) The *far jet* region where the studies assume that the jet enters the region with supersonic (or super fast magnetosonic) speed. (2) The *near jet* region where the matter enters with subsonic (or sub slow magnetosonic) speed and exits with supersonic speed. This is the jet formation region. (3) The *disk* region includes the problem of the Balbus-Hawley [39] and other instabilities [40] and the resulting 3-D magnetohydrodynamic (MHD) turbulence. There is of course a strong physical coupling between the disk and near jet region. Of particular interest is the back reaction of an MHD jet on the accretion rate [41], [42].

Many detailed hydro and magneto-hydrodynamic simulation studies have been done for the far jet region (see, for example, for nonrelativistic flows, [43], [44]; for relativistic hydro, Duncan and Hughes [45], [46]; and for relativistic MHD, van Putten [47]). Much less attention has been given to the MHD simulation of the near jet region [48]–[53]. A few simulations address the coupled problem of the disk and near jet regions [54], [55].

Here, we review recent results on nonrelativistic MHD simulations of the formation of jets. In Section 1, we discuss general considerations of MHD jet formation. In Section 2, we first discuss simulations of MHD jet formation which leads in one limit (Section 2.1) to nonstationary jet formation, and in a second limit (Section 2.2) to stationary jet formation. In Section 3, we discuss the opening of the magnetic field of a star which threads a surrounding accretion disk and the related possibility of MHD outflows. This situation is important to the case of X-ray pulsars in binary systems. Section 4 gives conclusions from this work.

1. BASICS OF MHD OUTFLOWS FROM DISKS

The main forces which are thought to drive matter to outflows from a rotating disk are thermal and magnetic pressure gradients, and the centrifugal force. The main opposing force is the gravity of the central object. Here, we do not consider radiatively driven outflows (see [56]). We consider a system consisting of a gravitating center (a star or black hole) and an accretion disk. We suppose that accretion disk is threaded by an ordered magnetic field and concentrate attention on the influence of the ordered magnetic field. Accreting matter of the disk carries magnetic field flux inward, thus generating a B_r component of the magnetic field as sketched in FIG. 2. On the other hand, rotation of the disk acts to generate a toroidal component of the field $B_\phi(< 0$ if $B_z > 0)$.

An accretion disk with a sufficiently inclined magnetic field (θ in FIG. 2 sufficiently large) can generate outflows as a result of the centrifugal force [16]–[18] or as a result of the magnetic pressure gradient force (appproximately $-\nabla_z B_\phi^2/(8\pi)$) [1], [36]. In spite of the enormous difference in scales between stellar jets (lengths ~1pc) and extragalactic jets (lengths ~1–10^3 kpc), the physics of the matter outflow from the disk may be the same and depends on the ratio of energy densities at the base of the outflow at the inner radius of the disk denoted r_i. Thus, the main parameters are

$$\varepsilon_{th} = (c_s/v_K)_i^2, \quad \varepsilon_B = (v_A/v_K)_i^2, \tag{1}$$

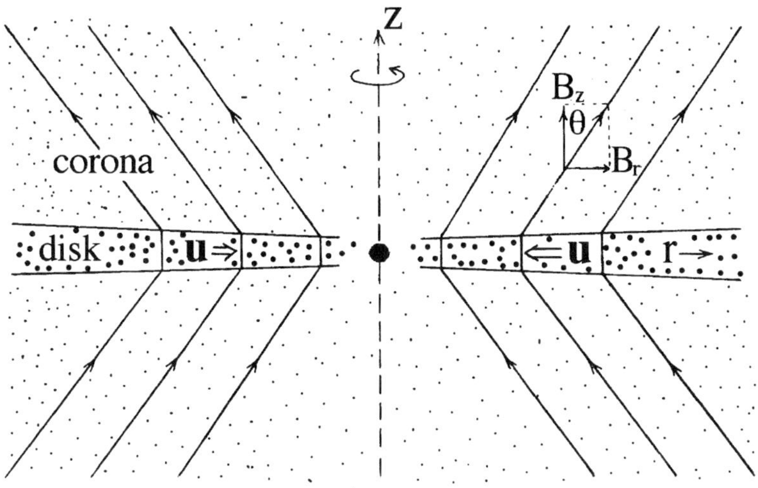

FIGURE 2. Sketch of a magnetized accretion disk with corona.

where v_K is Keplerian velocity, c_s the sound speed, v_A the Alfvén velocity, and the i subscript indicates evaluation at $r = r_i$ on the disk surface. For ε_B of order unity, the outflow is magnetically driven, whereas for ε_{th} of order unity, the flow is thermally driven.

As mentioned above, processes in the disk are coupled to the outflows (see also [57]). But it is difficult to simulate both regions simultaneously because the time scales of the accretion and outflow are in general very different. The accretion is much slower. On the other hand the processes in the disk involve small scale instabilities [39] and therefore require high spatial resolution. Stone and Norman [54] attempted to simultaneously simulate the internal MHD dynamics of a disk and MHD dynamics of outflows. This proved impractical because essentially all of the spatial resolution was needed for treating the unstable dynamics of the disk. Simulation of the internal MHD disk dynamics has led several groups to the problem of simulating 3-D MHD turbulence in a sheared flow (for example, [58], [59]). This is a much larger project than that of understanding MHD outflows. At the same time it is widely thought, and observations of cataclysmic variables support the view, that the disk turbulence — including MHD turbulence — can be modeled approximately using the Shakura-Sunyaev [60] alpha viscosity model [61], [62]. In contrast with the internal disk dynamics, there is theoretical and simulation evidence that the outflows can be treated using axisymmetric (2-D) MHD [17], [1], [51], [52]. Here, we consider outflows from a disk represented as a boundary condition (see also [53]). This corresponds to the outflows from a disk with accretion speed small compared with the Keplerian speed.

2. NUMERICAL SIMULATIONS OF MHD OUTFLOWS

In order to test the analytical models of stationary magnetically or centrifugally driven jets, we performed numerical simulations of MHD outflows from a disk treated as a boundary condition. The disk was threaded initially by a poloidal magnetic field inclined to the disk. We investigated two configurations of the magnetic field. The first configuration was a superposition of virtual monopoles located on the z-axis under the disk, and the second was a tapered virtual monopole also located under the disk.

2.1. Nonstationary Outflows

The initial field configuration was a superposition of monopoles chosen so that the magnetic field was significantly inclined to the disk ($\theta > 30°$) over most of the disk surface. Matter of the corona was initially in thermal equilibrium with the gravitating center. At $t = 0$ we started to rotate the disk with Keplerian velocity and at the same time to push matter from the disk with a small poloidal velocity equal to a fraction of the slow magnetosonic velocity ($v_p = \alpha v_{sm}$, with $\alpha = 0.1$–0.9). For the simulations presented here, $\varepsilon_B = 0.03$, and $\varepsilon_{th} = 0.9$. Matter, coming from the disk had the same temperature as matter in the corona. Thus, we generated outflows with relatively high temperature and low magnetic field (see details in [51] and [52]).

We found that matter outflowing from the disk is accelerated as a result of both thermal and magnetic pressure gradients, and that the outflow collimates very rapidly (after several rotation periods of the inner radius of the disk) as a result of the pinching force of the toroidal magnetic field. The behavior of the outflows is shown in FIG. 3. The maximum of the outflow from the disk was found to coincide with the maximum of the toroidal magnetic field. Analysis of the forces [52] shows that the main forces in the z-direction are thermal and gravitational forces, which compensate each other, whereas the magnetic force, in spite of its small initial value, determines the acceleration of matter. The results shown in FIG. 3 used the complete system of ideal nonrelativistic MHD equations assuming axisymmetry but taking into account all three components of velocity and magnetic field [51]. The boundary conditions on the walls and on the top of the cylinder were taken as free boundary conditions for all scalar quantities and components of vector quantities.

Analysis of the fluxes of mass, energy, and angular momentum through a horizontal surface, say, $z = 0.5\,z_{\max}$ shows that initally the fluxes grow as a result of the initial magnetic field twist formation due to the rotation of the disk. This twist of the field forms first at the inner radius of the disk where the disk rotation is fastest. Later the twist develops at larger radii. The matter flux grows and later decreases to a relatively small value. The velocity of matter on top of the region is twice the escape velocity during the maximum of the outflow, but becomes less than the escape velocity later when the mass flux decreases. Thus, this simulation is an example of a temporary outburst of matter to a jet. One objective in this study was to investigate outflows for an inclined magnetic field. However, the field rapidly becomes collimated — not inclined — as a result of the toroidal magnetic field. The case described in the next subsection was chosen so as to diminish the collimating effect of the toroidal magnetic field.

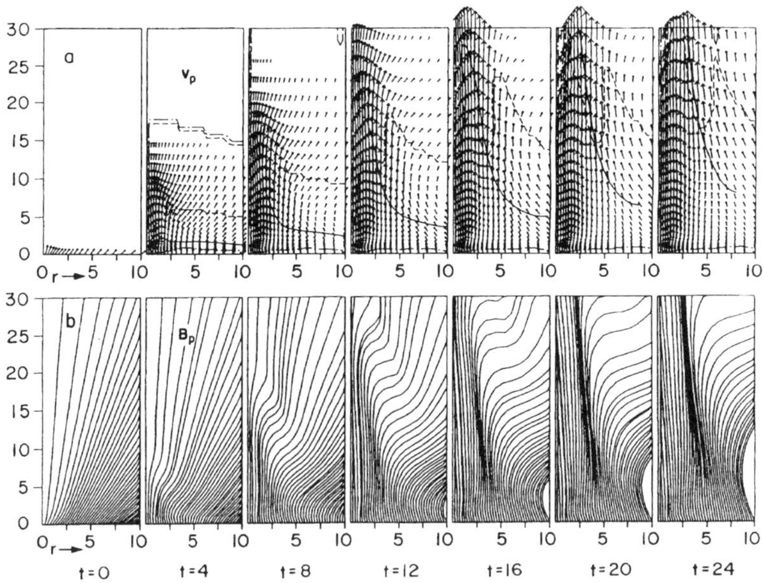

FIGURE 3. Outflow for a superposition of monopoles. **(a)** poloidal velocities of the flow; **(b)** poloidal magnetic field B_p at times $t = 0, 4, 8$, etc., where time t is measured in units of t_i which is the period of rotation of the inner radius of the disk (r_i). Only one quadrant of the flow is shown. The long dashed line (a) represents the slow magnetosonic surface; the short dashed line is the Alfvén surface; and the dashed line on top of the region is the fast magnetosonic surface. At the top of the region the flow exceeds the escape velocity.

2.2. Stationary Outflows

This simulation differed from the previous one in several respects (see [63]). First, we considered a different configuration of the initial poloidal magnetic field, namely, we took $B_p = B_M/(1 + \eta \tan^2\theta)$, where B_M is the monopole magnetic field,θ is an angle between the magnetic field line and the z-axis, and η = constant (typically, 1–10). This field is close to a monopole in the inner part of the disk and near the z-axis, and it goes to zero in the outer part of the disk. We chose this configuration in order to decrease the collimating affect of the toroidal magnetic field. A weak poloidal magnetic field will give rise to a weak toroidal magnetic field in the outer regions of the flow (see [1]), and this will decrease the collimation.

Second, we considered outflows at different values of ε_B including the case of strong magnetic fields with $\varepsilon_B = 1$, and different temperatures ε_{th}. In contrast with the previous case, we supposed that the entropy is constant everywhere so that the energy equation is not needed. This was done taking into account that we are interested mainly in the stong magnetic field case where the thermal pressure is small compared with magnetic pressure.

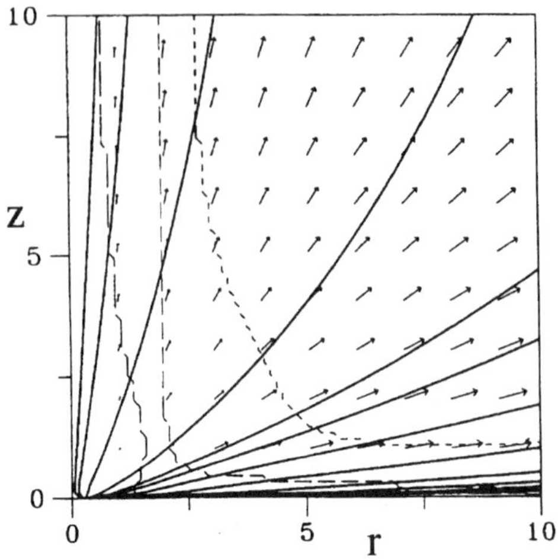

FIGURE 4. Outflow in the case of a tapered monopole field for $\varepsilon_B = 0.81$, $\varepsilon_{th} = 0.25$, $\eta = 2$ at time $t = 100t_i$, when a stationary flow is well established, where t_i is the period of rotation at the inner radius of the disk (r_i). The solid lines are flux lines and show the poloidal magnetic field. The arrows are the poloidal velocities. The long dashed line is the slow magnetosonic surface; the dashed line is the Alfvén surface; and the short dashed line is the fast magnetosonic surface. At the outer part of the region the flow exceeds the escape velocity.

After about 40 rotation periods, we found stationary outflows as shown in FIG. 4. Matter from the disk goes through the slow magnetosonic, Alfvén, and fast magnetosonic surfaces. The fluxes of mass, energy, and angular momentum through a horizontal surface, say, $z = 0.5\,z_{max}$ grow and after about 40 rotation periods of the inner disk they approach steady values. It is of interest to note that in many simulation runs the energy flow due to the electromagnetic fields is larger than that carried by the matter.

In the stationary limit, the field lines leaving the disk are strongly inclined away from the jet axis with the angle $\theta \sim 70$–$75°$. Within the computational region the collimation is weak. However, collimation can occur at larger distances owing to the pressure of an external medium [1]. Analysis of the forces along poloidal magnetic field lines shows that in most of the region (excluding the region with small θ) the main driving force is the magnetic pressure gradient, approximately $-\nabla B_\phi^2/(8\pi)$. Thus, these solutions are essentially in accord with the magnetically driven outflows of Lovelace *et al.* [1].

This type of field configuration may arise under conditions where the magnetic field threading the disk has different polarities at different disk radii (say, positive for r_1 to r_2, and negative for r_2 to r_3, etc.). Such polarity variations are expected, for example, in AGN disks for a **B** field random over large scales accreted from the interstellar medium or a large scale ($\Delta r/r \sim 1$) random field due to a disk dynamo (see [42]). At the radii in the disk where the poloidal field changes signs, the magnetic collimation is small. The polarity reversals in the disk will produce jets which at large distances have polarity reversals of **B** as a function of z. In the vicinity of the field reversals, collisionless reconnection may be an important particle acceleration mechanism (see [64]).

3. OPENING OF MAGNETIC FIELD LINES AND OUTFLOWS

Another important **B** field configuration relevant to MHD outflows from a disk is that of a dipole field of a star which initially threads a surrounding accretion disk. The star may be an accreting pulsar (for example, in a binary X-ray source) or a young stellar object. The question is: What is the resultant field configuration taking into account the difference in the angular rotation rates of the foot points of field lines? Newman *et al.* [65] studied the evolution of a closed, axisymmetric, force-free coronal magnetic field which threads a Keplerian disk at different radii and showed that the difference in angular velocities of the foot points leads to the formation of toroidal magnetic field, growth of the coronal magnetic field energy, and ultimately to opening of the magnetic field lines (see also [66] and [67]). Lovelace *et al.* [68] studied the system consisting of a rotating star with an aligned dipole field and an accretion disk — a situation discussed earlier by Ghosh and Lamb [69]. They concluded that in an inner region, the magnetic field remains attached to the star forming a closed magnetosphere, while in the exterior region, opened magnetic field lines thread the accretion disk as sketched in FIG. 5. Under some conditions, the open magnetic field region of the disk can lead to the formation of MHD outflows [68].

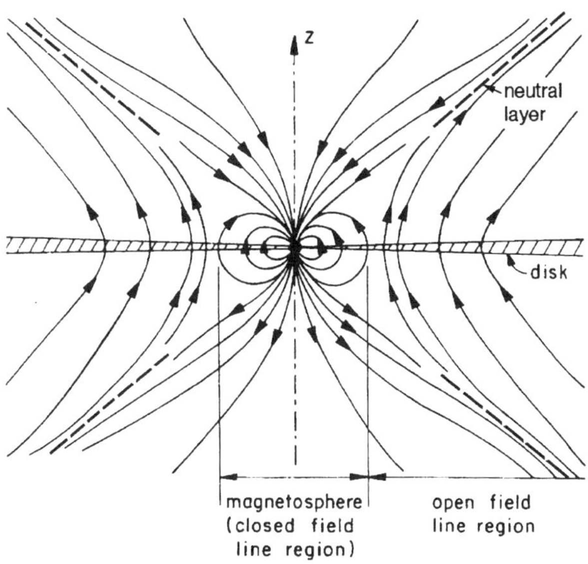

FIGURE 5. Schematic drawing of the poloidal magnetic field configuration in a system consisting of a star with an aligned dipole magnetic field and a differentially rotating accretion disk. MHD outflows may occur from the open field line region of the disk. The figure is adapted from [68].

The opening of an initially closed poloidal magnetic field configuration linking a star and a disk has been observed in numerical simulations by Miller and Stone [55] and Romanova *et al.* [70]. They found opening of the field lines and strong magnetically driven outflows in the case where the plasma is highly conducting. Miller and Stone noted, however, the possibility of a closed magnetosphere extending out to large distances from the star in the case considered by Ghosh and Lamb [69] where the conductivity is anomalously low.

4. CONCLUSIONS

The simulation results discussed here support the idea that MHD processes lead the formation of jets near the inner edges of accretion disks, and that the jets have asymptotic speeds of the order of the maximum Keplerian velocity of the disk. In contrast, observed VLBI jets in quasars and active galaxies point to bulk Lorentz factors of order 10 — much larger than the disk Lorentz factor. This may be a result of the relativistic dynamics not included here, but more likely it reflects the fact that these jets are in a different regime — possibly the Poynting flux regime discussed by Lovelace et al. [11] — and/or that the jets involve energy extraction from a rotating black hole [71].

REFERENCES

1. LOVELACE, R.V.E., H.L. BERK & J. CONTOPOULOS. 1991. Magnetically driven jets and winds. Astrophys. J. **379:** 696–705.
2. BRIDLE, A.H. & J.A. EILEK, EDS. 1984. Physics of Energy Transport in Extragalactic Radio Sources, Greenbank: NRAO.
3. MIRABEL, I.F. & L.F. RODRIGUEZ. 1994. A superluminal source in Galaxy. Nature. **371:** 46–48
4. MUNDT, R. 1985. Highly collimated mass outflows from young stars. *In* Protostars and Planets II. D.C. Black & M.S. Mathews, Eds.: 414–433. Univ. of Arizona Press. Tucson, AZ.
5. BÜHRKE, T., R. MUNDT & T.P. RAY. 1988. A detailed study of HH 34 and its associated jet. Astron. & Astrophys. **200:** 99–119.
6. BEGELMAN, M.C., R.D. BLANDFORD & M.J. REES. 1984. Theory of extragalactic radio sources. Rev. Mod. Phys. **56:** 255–351.
7. BISNOVATYI-KOGAN, G.S. 1993. Mechanisms of jet formation. *In* Stellar Jets and Bipolar Outflows. L. Errico & A.A. Vittone, Eds.: 369–381. Kluwer. Dordrecht, The Netherlands.
8. GOLDREICH, P. & W. JULIAN. 1969. Pulsar electrodynamics. Astrophys. J. **157:** 869–880.
9. LOVELACE, R.V.E. 1976. Dynamo model of double radio sources. Nature. **262:** 649–652.
10. BLANDFORD, R.D. 1976. Accretion disk electrodynamics — a model for double radio sources. Mon. Not. Roy. Astron. Soc. **1976:** 465–481.
11. LOVELACE, R.V.E., J.C.L.WANG & M.E. SULKANEN. 1987. Self-collimated electromagnetic jets from magnetized accretion disks. Astrophys. J. **315:** 504–535.
12. GISLER, G., R.V.E. LOVELACE & M.L. NORMAN. 1989. Electrodynamic formation of astrophysical jets: Simulations. Astrophys. J. **342:** 135–145.
13. WANG, J.C.L., R.V.E. LOVELACE & M.E. SULKANEN. 1990. Self-collimated electromagnetic jets from magnetized accretion disks — the even symmetry case. Astrophys. J. **353:** 38–43.

14. WEBER, E.J. & L. DAVIS, JR. 1967. The angular momentum of solar wind. Astrophys. J. **148:** 217–227.
15. MESTEL, L. 1968. Magnetic braking by a stellar wind. Mon. Not. Roy. Astron. Soc. **138:** 359–391.
16. KÖNIGL, A. & S.P. RUDEN. 1993. Origin of Outflows, E.H. Levy & J. Lunine, Eds.: 641–687. Univ. of Arizona Press. Tucson, AZ.
17. BLANDFORD, R.D. & D.G. PAYNE. 1982. Hydrodynamic flows from accretion disks and the production of radio jets. Mon. Not. Roy. Astron. Soc. **199:** 883–903.
18. PUDRITZ, R.E. & C.A. NORMAN. 1986. Bipolar hydrodynamic winds from disks around protostellar objects. Astrophys. J. **301:** 571–586.
19. KÖNIGL, A. 1989. Self-similar models of magnetized accretion disks. Astrophys. J. **342:** 208–223.
20. PELLETIER, G. & R.E. PUDRITZ. 1992. Hydrodynamic disk winds in young stellar objects and active galactic nuclei. Astrophys. J. **394:** 117–138.
21. CONTOPOULOS, J. & R.V.E. LOVELACE. 1994. Magnetically driven jets and winds: Exact solutions. Astrophys. J. **429:** 139–152.
22. SHU, F.H., S. LIZANO, S.P. RUDEN & J. NAJITA. 1988. Mass loss from rapidly rotating magnetic protostars. Astrophys. J. **328:** L19– L23.
23. PRINGLE, J.E. 1989. A boundary layer origin for bipolar flows. Mon. Not. Roy. Astron. Soc. **236:** 107–115.
24. ZEHRFELD, H.P. & B.J. GREEN. 1972. Stationary toriodal equilibria at finite beta. Nucl. Fusion. **12:** 569–575.
25. HEINEMANN, M. & S. OLBERT. 1978. Axisymmetric ideal MHD stellar wind flow. J. Geophys. Res. **83:** 2457–2460.
26. ALY, J.J. 1986. Some topics in the magnetohydrodynamics of accreting magnetic objects. *In* Magnetospheric Phenomena in Astrophysics. AIP Conference Proceedings No. 144. R.I. Epstein & W.C. Feldman, Eds.: 45–123. Am. Inst. of Physics. New York.
27. SAKURAI, T. 1987. Magnetically collimated winds from accretion disks. Pub. Astron. Soc. Jap. **39:** 821–835.
28. LOVELACE, R.V.E., C. MEHANIAN, C.M. MOBARRY & M.E. SULKANEN. 1986. Theory of axisymmetric magnetohydrodynamic flows: Disks. Astrophys. J. Suppl. **62:** 1–37.
29. CAMENZIND, M. 1986. Hydromagnetic flows from rapidly rotating compact objects. Acta Astron. **162:** 32–44.
30. TSINGANOS, K., C. SAUTY, G. SURLANTZIS, E. TRUSSONI & J. CONTOPOULOS. 1996. Critical points and separatrix characteristics in solar and astrophysical MHD flows. *In* Solar and Astrophysical Magnetohydrodynamic Flows. K.C. Tsinganos, Ed.: 427–458. Kluwer Acad. Pub. Dordrecht,The Netherlands.
31. GRAD, H. & H. RUBIN. 1958. Hydromagnetic equilibria and force-free fields. *In* Proc. Internat. Atomic Energy Agency Conf., IAEA, Geneva, 31, p. 190.
32. SHAFRANOV, V.D. 1966. Plasma equilibrium in a magnetic field. Rev. of Plasma Phys. **2:** 103–151.
33. CHANDRASEKHAR, S. 1956. Axisymmetric magnetic fields and fluid motions. Astrophys. J. **124:** 232–243.
34. WOLTJER, L. 1959. Hydromagnetic equilibrium. IV. Axiymmetric compressible media. Astrophys. J. **130:** 405–413.
35. MESTEL, L. 1961. A note on equatorial acceleration in a magnetic star. Mon. Not. Roy. Astron. Soc. **122:** 473–478.
36. KOUPELIS, T. & H.M. VAN HORN. 1989. A quasi-one-dimensional model for narrow astrophysical jets. Astrophys. J. **342:** 146–186.

37. Lovelace, R.V.E., C.M. Mobarry & J. Contopoulos. 1989. Magnetohydrodynamic flows in accretion disks and jets. *In* Accretion Disks and Magnetic Fields in Astrophysics, G. Belvedere, Ed.: 71–90. Kluwer. Dordrecht, The Netherlands.
38. Lovelace, R.V.E., M.M. Romanova & J. Contopoulos. 1993. Theory of jets from young stars. Astrophys. J. **403:** 158–163.
39. Balbus, S.A. & J.F. Hawley. 1991. A powerful local shear instability in weakly magnetized disks. I. Linear analysis. Astrophys. J. **376:** 214–222.
40. Kaisig, M., T. Tajima & R.V.E. Lovelace. 1992. Magnetic interchange instability of accretion disks. Astrophys. J. **386:** 83–89.
41. Lovelace, R.V.E., M.M. Romanova & W.I. Newman. 1994. Implosive accretion and outbursts of active galactic nuclei. Astrophys. J. **437:** 136–143.
42. Lovelace, R.V.E., W.I. Newman & M.M. Romanova. 1997. Jet outbursts from fast accretion in a disk with zebra-stripe magnetic field. Astrophys. J. **484:** 628–636.
43. Clarke, D.A., M.L. Norman & J.O. Burns. 1986. Numerical simulations of magnetically confined jet. Astrophys. J. 3**11:** L63–L67.
44. Hardee, P.E. & D.A. Clarke. 1992. Nonlinear dynamics of three-dimensional jet. Astrophys. J. **400:** L9–L12.
45. Duncan, G.C. & P.A. Hughes. 1994. Simulations of relativistic extragalactic jets. Astrophys. J. **436:** L119–L122
46. Gomez, J.L., J.M. Marti, A.P. Marscher, J.M. Ibanez, J.M. Marcaide. 1995. Parsec-scale synchrotron emission from hydrodynamic relativistic jets in active galactic nuclei. Astrophys. J. **449:** L19–L21.
47. van Putten, M.H.P.M. 1993. A two-dimensional relativistic ($\Gamma = 3.25$) jet simulation. Astrophys. J. **408:** L21–L23.
48. Uchida,Y & K. Shibata. 1985. Magnetodynamical acceleration of CO and optical bipolar flows from the region of star formation. Pub. Astron. Soc. Jap. **37:** 515–535.
49. Shibata, K. & Y. Uchida. 1986. A magnetodynamic mechanism for the formation of astrophysical jets. I. Dynamical effects of the relaxation of nonlinear magnetic twists. Pub. Astron. Soc. Jap. **37:** 31–46; A magnetodynamic mechanism for the formation of astrophysical jets. II. Dynamical processes in the accretion of magnetized mass in rotation. Pub. Astron. Soc. Jap. **38:** 631–660.
50. Bell, A.R. 1994. Magnetohydrodynamic jets. Phys. Plasmas. **1:** 1643–1652.
51. Ustyugova, G.V., A.V. Koldoba, M.M. Romanova, V.M. Chechetkin & R.V.E. Lovelace. 1995. Magnetohydrodynamic simulations of outflows from accretion disks. Astrophys. J. **439:** L39–L42.
52. Koldoba, A.V., G.V. Ustyugova, M.M. Romanova, V.M. Chechetkin & R.V.E. Lovelace. 1995. Simulations of jet formation from a magnetized accretion disk. Astrophs. & Spa. Sci. **232:** 241–261.
53. Meier, D.L., D.G. Payne & K.R. Lind. 1995. A magnetic switch that determines the speed of astrophysical jets. Nature **388:** 350–352.
54. Stone, J.M. & M.L. Norman. 1994. Numerical simulations of magnetic accretion disks. Astrophys. J. **433:** 746–756.
55. Miller, K.A. & J.M. Stone. 1997. MHD simulations of stellar magnetosphere-accretion disk interaction. Preprint.
56. Phinney, E.S. 1987. How fast can a blob go? *In* Superluminal Radio Sources, J.A. Zensus & T.J. Pearson, Eds.: 301–305. Cambridge Univ. Press. Cambridge.
57. Falcke, H., M.A. Malkan & P.L. Biermann. 1995. The jet-disk symbiosis. II. Interpreting the radio/uv correlations in quasars. Astron. & Astrohys. **298:** 375–394.

58. HAWLEY, J.F., C.F. GAMMIE & S.A. BALBUS. 1995. Local three-dimensional magnetohydrodynamic simulations of accretion disks. Astrophys. J. **440:** 742–763.
59. BRANDENBURG, A., A. NORDLUND, R.F. STEIN & U. TORKELSON. 1995. Dynamo generated turbulence and large-scale magnetic fields in a Keplerian shear flow. Astrophys. J. **446:** 741–754.
60. SHAKURA, N.I. & R.A. SUNYAEV. 1973. Black holes in binary systems. Observational appearance. Astron. & Astrophys. . **24:** 337–355.
61. EARDLEY, D.M. & A.P. LIGHTMAN. 1975. Magnetic viscosity in relativistic accretion disks. Astrophys. J. **200:** 187–203.
62. CORONITI, E.V. 1981. On the magnetic viscosity in Keplerian accretion disks. Astrophys. J. **244:** 587–599.
63. ROMANOVA, M.M., G.V. USTYUGOVA, A.V. KOLDOBA, V.M. CHECHETKIN & R.V.E. LOVELACE. 1997. Formation of stationary magnetohydrodynamic outflows from a disk by time-dependent simulations. Astrophys. J. **482:** 708–711.
64. ROMANOVA, M.M. & R.V.E. LOVELACE. 1992. Magnetic field, reconnection, and particle acceleration in extragalactic jets. Astron. & Astrophys. **262:** 26–36.
65. NEWMAN, W.I., A.L. NEWMAN & R.V.E. LOVELACE. 1992. Initiation of bipolar outflows by magnetic field twisting in protostellar nebulae. Astrophys. J. **392:** 622–627.
66. ALY, J.J. 1984. On some properties of force-free magnetic fields in infinite regions of space. Astrophys. J. **283:** 349–362.
67. LYNDEN-BELL, D. & C. BOILY. 1994. Self-similar solutions up to flashpoint in highly wound magnetostatics. Mon. Not. Roy. Astron. Soc. **267:** 146–152.
68. LOVELACE, R.V.E., M.M. ROMANOVA & G.S. BISNOVATYI-KOGAN. 1995. Spin-up/spin-down of magnetized stars with accretion disks and outflows. Mon. Not. Roy. Astron. Soc. **275:** 244–254.
69. GHOSH, P. & F.K. LAMB. 1978. Disk accretion by magnetic neutron stars. Astrophys. J. **223:** L83–L87.
70. USTYUGOVA, G.V., A.V. KOLDOBA, M.M. ROMANOVA, V.M. CHECHETKIN & R.V.E. LOVELACE. 1997. Dynamics of magnetic loops in the coronae of accretion disks. Astrophys. J. June 20.
71. BLANDFORD, R.D. & R.L. ZNAJEK. 1977. Electromagnetic extraction of energy from Kerr black holes. Mon. Not. Roy. Astron. Soc. **179:** 433–456.

Buoyancy-Driven Rotating Boundary Currents[a]

P.A. YECKO[b] AND S.P. MEACHAM[c]

[b]*Physics Department, University of Florida, Gainesville, Florida 32611*
[c]*Department of Oceanography, Florida State University, Tallahassee, Florida 32306*

ABSTRACT: The structure of boundary currents formed from intermediately dense water introduced into a rotating, stably stratified, two-layer environment is investigated in a series of laboratory experiments, performed for Froude numbers ranging from 0.01 to 1. The thickness and streamwise velocity profiles in quasi-steady currents are measured and found to agree with analytic solutions and numerical models. A sloping boundary in a stratified background results in robust stability at all experimental Froude numbers. Such stability is in sharp contrast to the rapid breakup into eddies of such a current flowing against a vertical boundary, or of a current flowing along a slope in a uniform background. The possibility of wave radiation supported by a variety of wave mechanisms found in the ambient fluid might account for the observed stability.

1. INTRODUCTION

Buoyancy is responsible for the formation of many important geophysical and astrophysical flows. In a rotating system, a sustained source of positive or negative buoyancy may give rise to an extended horizontal current against a lateral boundary, a gravity current. In such a current, the Coriolis force associated with the horizontal flow in the current is balanced by an offshore pressure gradient supported by the wall. The resulting flow is then bounded by a solid boundary on the right (looking along the direction of flow in the Northern hemisphere), and by a front on the left. If the wall is sloping rather than purely vertical, the component of the Coriolis force parallel to the wall in the vertical plane can balance the similar component of the gravitational force acting on the current. Where the source is persistent, such currents can reach states of quasi-equilibrium. See Griffiths [1] for a review of boundary gravity currents.

It is known that the instability of many varieties of sheared flows cannot be determined locally [2]. As an illustrative example, recall that for a two-dimensional incompressible shear flow, $U = U(y)$, one *must* use boundary conditions appropriate to the corresponding eigenvalue problem in order to obtain a necessary condition for instability. That condition is simply the vanishing of the vorticity gradient ($U_{yy} = 0$) somewhere in the flow — the celebrated Rayleigh criterion. Local approximation leads to a different condition and the erroneous prediction of likely instability [3].

Investigation of more complicated (compressible) 2-D shear flows has confirmed the nonlocality of wave-interaction, or resonant, instabilities [4] and more specific studies of boundary current configurations which account for the coupling of the

[a]This work was supported by the National Science Foundation as part of contract OCE-9301318.

current to its environment have also found that mode resonances can alter the intrinsic stability characteristics of flows [5]–[7].

2. THE MEDDY PROBLEM

How can the presence of warm, salty Mediterranean water be explained when it is found in the western Atlantic ocean? The Mediterranean water is delivered in compact lens-shaped vortices affectionately known as *Meddies*. Such long-range mixing capabilities of some currents has a profound influence on the global ocean circulation. This example poses two distinct questions:

1. How are Meddies formed?
2. How do they subsequently propagate?

The propagation of coherent patches of vorticity is only partially understood. A number of different long-range mechanisms can affect the motion of vortices. These include: interactions between different vortices, advection by a large scale background flow, interaction with a background vorticity or potential vorticity gradient (a good example of this is the "planetary beta" effect seen in flows in a thin shell of fluid on the surface of a rotating spherical planet [8]), interaction of a vortex with a horizontal or vertical boundary, small scale dissipation within and around a vortex, and MHD effects in examples of astrophysical vortices.

3. LABORATORY EXPERIMENTS

To investigate Question 1, we developed a series of laboratory experiments to study buoyancy-driven currents in rotating stratified environments. Although we have targeted a specific oceanographic flow to model, it is tantalizing to consider that a reduced problem such as the one we propose here may also be relevant to other problems. In binary star systems, to consider one possible example, infall material from one star can feed an accretion disk around the other; the subsequent equilibration of the new material may well lead to buoyancy driven currents of a similar nature. But this study is primarily a model for the Mediterranean outflow, where we have the benefit of comparison with direct observations. The two principal goals of the experiment were:

(a) to determine the structure (velocity shear and thickness);
(b) to examine the stability properties of these currents.

A diagram of the experimental apparatus is shown in FIG. 1. The interface of the stable ambient two-layer stratification intersects the surface of a conical section at mid-depth. A thymol blue solution of intermediate density is held outside the tank and also fills a reservoir at the tank's outer edge. The densities of the overflow water (ρ_2) and the bottom layer (ρ_3) were achieved by dissolving sugar in distilled water.

A current is initiated by establishing a flow that causes overflow in the reservoir and the subsequent buoyant descent of the overflow water down the sloping surface toward the layer interface, where it becomes neutrally buoyant. Because of the rotation, the descending current turns to flow cyclonically along the topographic slope.

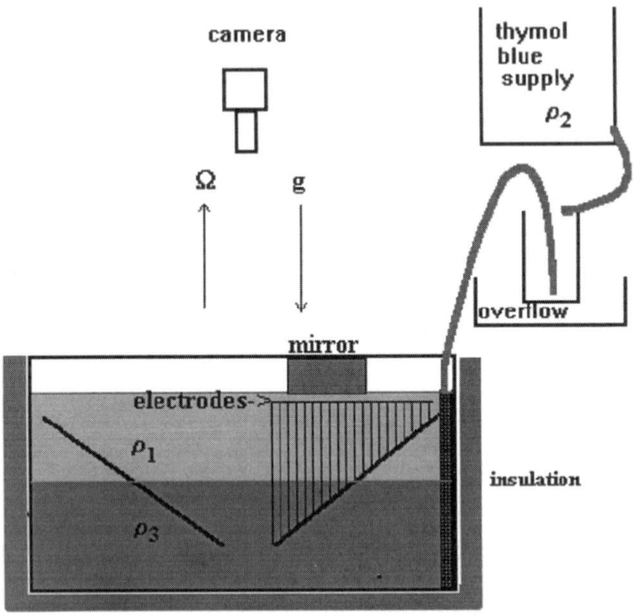

FIGURE 1. The experimental apparatus: the cylindrical tank and the submerged conic slope are shown in cross-section; the tank has a diameter of ~60 cm. Everything sits on a rotating table.

Except for extreme initial conditions, the current will still reach the layer interface. Along one radius of the conical surface, an array of fine (0.006″ diameter) vertical wires are placed at 0.44 cm intervals. The wires are pulsed with an electrical current to activate thymol blue tracer in fluid of the current, allowing direct visualization. A camera is positioned above the tank (where the length of advected tracer is used to determine velocities) and a mirror is oriented to allow a head-on view of the current (where the height of unadvected tracer determines local thickness). Crude evaluation of the response in the ambient layers is done by tracking floating and, where possible, submerged tracers. Images captured with the CCD camera were digitally stored using a video framer and the measurement of tracer was facilitated by enhancing contrast using standard algorithms. Pre-marked intervals on the experimental surfaces provided accurate distance standards, and the real-time record kept on video simplified measurement of time intervals.

In all experiments, the densities were prepared with the values: $\rho_1 = 1.0$ g/cc, $\rho_2 = 1.005$ g/cc, and $\rho_3 = 1.01$ g/cc; the depths of the ambient layers were 11 cm each. The reduced gravity between layers then took the values: $g_{12}' = g_{23}' = 5.0$ g cm/sec^2, while Ω was either 1. or 1.5. The principal control value was the volume flux, S, of ρ_2-fluid, which was varied from 0.15–1.0 cc/sec. S is directly related to a character-

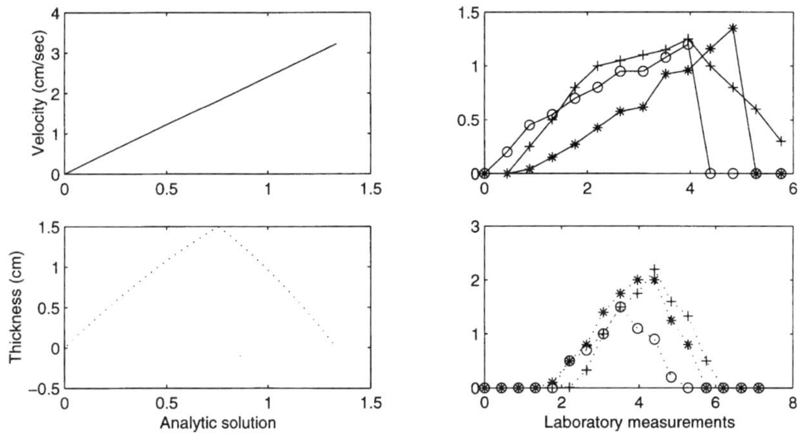

FIGURE 2. **(Left)** Steady solutions for u-velocity and thickness of the boundary current in the three-layer system flowing along a sloping boundary. **(Right)** The measured u-velocities and thickness for three experiments, one case conincides with the shedding of a cyclone.

istic current depth H, which varied from 1–3 cm. Two values of the topographic slope, α, were used, 0.0 (a vertical wall) and 0.25.

From the parameters above, we can construct several useful quantities. A Rossby deformation length is given by $L_D = (g'H)^{1/2}/2\Omega$, and the ratio of the current width to L_D is indicative of the relative importance of baroclinic and barotropic disturbances. A Reynolds number is given by $Re = (g'H^3/\nu^2)^{1/2}$, and for these experiments reached nearly 10^3, but a more useful quantity is the Froude number $F = U^2/(g'H)$.

These and other experiments have verified that baroclinic currents and instabilities of interest are characterized not so much by Re as by F. In oceanic flows, F can vary between 10^{-2} and 10^2, and in the laboratory we can easily produce a range of F between 10^{-2} and 10, simply by forming velocities U in the range 0.1 to 1 cm/sec.

In high Reynolds number flows without ambient stratification, experiments have shown that instabilities of Kelvin-Helmholtz type typically grow rapidly at the nose, while further upstream — where the steady current is well established — baroclinic instabilities produce eddies. We focus on the established current rather than its nose and the details of the intrusion process. For this reason, we initiate the current with a small density enhancement to minimize the initial Kelvin-Helmholtz mixing, while using a tank of sufficient horizontal dimension to traverse the relevant Froude number range.

4. STEADY THREE LAYER CURRENTS

What is the three-layer, steady, streamwise uniform solution? An arbitrary three layer arrangement of fluid is described by the familiar shallow layer system:

$$\frac{d\mathbf{u}_1}{dt} + \mathbf{f} \times \mathbf{u}_1 = -g\nabla(h_1 + h_2 + h_3), \tag{1}$$

$$\frac{d\mathbf{u}_2}{dt} + \mathbf{f} \times \mathbf{u}_2 = -g\frac{\rho_1}{\rho_2}\nabla h_1 - g\nabla(h_2 + h_3), \tag{2}$$

$$\frac{d\mathbf{u}_3}{dt} + \mathbf{f} \times \mathbf{u}_3 = -g\frac{\rho_1}{\rho_3}\nabla h_1 - g\frac{\rho_2}{\rho_3}\nabla h_2 + g\nabla h_3, \tag{3}$$

and

$$\frac{\partial h_i}{\partial t} + \nabla \cdot (h_i \mathbf{u}_i) = 0 \tag{4}$$

holds for each layer $i = 1, 2, 3$.

We will let Ox be the streamwise direction, parallel to the boundary, and Oy, the cross-stream direction. For a boundary current of intermediate density ($i = 2$) flowing against a vertical wall, a steady, streamwise uniform current with velocity $\bar{u}(y)$ and thickness $\bar{h}(y)$ must satisfy potential vorticity and momentum conservation:

$$\frac{d}{dt}\left(\frac{f - \partial_y \bar{u}}{\bar{h}}\right) = 0 \quad \text{and} \quad f\bar{u} = -g^* \partial_y \bar{h}, \tag{5}$$

where the reduced gravity here is

$$g^* = g\frac{(\rho_2 - \rho_1)(\rho_3 - \rho_2)}{\rho_2(\rho_3 - \rho_1)},$$

For uniform *nonzero* initial potential vorticity, $Q_2(t = 0) = Q_*$, we find currents of width L having:

$$\bar{u}(y) = fl\frac{\sinh(y/l)}{\cosh(L/l)} \quad \text{and} \quad \bar{h}(y) = h_* - h_*\frac{\cosh(y/l)}{\cosh(L/l)}, \tag{6}$$

where a deformation length is now associated with the initial conditions:

$$l = \left(\frac{g^*}{fQ_*}\right)^{1/2}. \tag{7}$$

To extend the solutions above to the laboratory configuration, in which the boundary along which the current flows has some slope, we must simply match the solution above to a two layer solution of a current flowing along a slope (this subproblem is the subject of several previous studies, see the review [1] for references).

5. RESULTS

Steady solutions are shown in FIG. 2 alongside measured profiles. Experiments performed with vertical boundaries have found the strong tendency of currents to be unstable, even when the currents were initiated from a uniform potential vorticity source. This instability set in uniformly along the current once it became sufficiently wide, forming a chain of eddies. The eddies were roughly the same width as the current and moved irregularly following separation.

Analogous experiments along sloping boundaries have found the currents to be stable over much longer times and for wider currents. The stabilization of such currents by the presence of a slope should be emphasized — it suggests that the instabilities found for highly idealized and often symmetrical posed problems may not be realized in more general settings. The first step in proving or disproving this hypothesis is understanding the steady current configurations and a subsequent stability analysis. What we have reported here are first measurements of steady current profiles for these stable cases.

Related experiments, in which a surface current is allowed to eventually encounter a sloping boundary, have also found that a slope can have a stablizing effect — following the encounter, the current was found to be slower, wider, and more laminar than it would be along a vertical wall [9].

A final point concerns the formation of Meddies: in the process of initiating experimental currents, the intermediate density fluid formed two distinct equilibria. The first as it flowed through less dense water, down the slope and to the right; after reaching the layer interface with the densest water, it spread to the right and along the sloping boundary. Both stages were rather laminar and steady, but at the transition, a single large cyclone was typically shed. The Mediterranean undergoes a similar transition where it negotiates a sharp turn at Cape St. Vincent and presumably other significant topographic variations, such as channels, are also encountered. The role of such localized perturbations may help explain why, when experimental currents suggest stability, their oceanic counterparts are able to form eddies.

The excitation of inertial waves in the ambient homogeneous fluid is also found.

Finally, we would like to use these preliminary results to initiate a more thorough investigation of the linear instability in the full three-layer problem as well as the basis for a companion set of numerical experiments to duplicate those done in the laboratory.

ACKNOWLEDGMENTS

The authors would like to thank Dr. Ruby Krishnamurti for being of invaluable assistance in the lab and thank the Geophysical Fluid Dynamics Institute at FSU for support and hospitality.

REFERENCES

1. GRIFFITHS, R.W. 1986. Gravity currents in rotating systems. Ann. Rev. Fluid Mech. **18:** 59–89.

2. KNOBLOCH, E. 1992. On the stability of magnetized accretion disks. Mon. Not. Royal Astr. Soc. **255:** 25p–28p.
3. BALBUS, S.A. & J.F. HAWLEY. 1991. A powerful local shear instability in weakly magnetized disks. I. Linear analysis. Ap. J. **376:** 214–233.
4. KNESSL, C. & J. KELLER. Stability of rotating shear flows in shallow water. J. Fluid Mech. **244:** 605–614.
5. KILLWORTH, P., N. PALDOR & M.E. STERN. 1984. Wave propagation and growth on a surface front in a two-layer geostrophic current. J. Marine Res. **42:** 761–785.
6. HAYASHI Y.-Y. & W.R. YOUNG. 1987. Stable and unstable shear modes of rotating parallel flows in shallow water. J. Fluid Mech. **184:** 477–504.
7. STEPHENS, J. AND S.P. MEACHAM. 1997. The stability of gravity currents along a slope. J. Phys. Oceanog. Submitted..
8. PEDLOSKY, J. 1987. Geophysical Fluid Dynamics. Springer-Verlag. Berlin.
9. WHITEHEAD, J. A. & D.C. CHAPMAN. 1986. Laboratory observations of a gravity current on a sloping bottom: The generation of shelf waves. J. Fluid Mech. **172:** 373.

Coherent Structures in Convection and Parametrically Driven Surface Waves[a]

JORGE VIÑALS

Supercomputer Computations Research Institute, Florida State University, Tallahassee, Florida 32306-4130 and Department of Chemical Engineering, FAMU/FSU College of Engineering, Tallahassee, Florida 32310-6046

ABSTRACT: A system maintained outside of thermodynamic equilibrium often develops complex spatial structures and, in some cases, persistent dynamics. The amplitude equation formalism has been widely used to study pattern formation in the weakly nonlinear regime (slightly beyond onset of instability from some stationary reference state.) Application of the formalism is briefly discussed for Rayleigh-Benard convection, and at some length for parametrically driven surface waves (Faraday waves).

1. INTRODUCTION

To understand the emergence of coherent structures in a system that is driven outside of thermodynamic equilibrium, one must begin with a stability analysis of a reference stationary state that becomes marginally stable for some particular value of an experimentally set control parameter. Let $\psi(x)$ be a particular state variable of a one-dimensional system, and ε a dimensionless control parameter: at $\varepsilon = 0$ the base state $\psi = 0$ loses stability against infinitesimal perturbations of the form,

$$\psi(x, t) = \psi_q(t)\, e^{iqx}, \tag{1}$$

for some $q = q_0$, the critical mode. Just above threshold ($\varepsilon > 0$), the neutral stability curve $\varepsilon = \varepsilon_c(q)$ (see FIG. 1a) is defined as the locus of wavenumbers q that are neutrally stable at ε. In this region, the growth rate of any unstable mode $\psi_q(t)$ is given by its distance away from the neutral curve,

$$\frac{d\psi_q}{dt} = [\varepsilon - \xi_0^2(q - q_0)^2]\psi_q + \dots, \tag{2}$$

where we have introduced a Taylor series expansion of the neutral stability curve around the critical wavenumber and defined,

[a]This research was supported by the U.S. Department of Energy, contract No. DE-FG05-95ER14566, and also in part by the Supercomputer Computations Research Institute, which is partially funded by the U.S. Department of Energy, contract No. DE-FC05-85ER25000.

$$\xi_0^2 = \frac{1}{2}\left(\frac{d^2\varepsilon_c(q)}{dq^2}\right)_{q = q_0}. \tag{3}$$

The quantity ξ_0 plays the role of a coherence length, and determines the generalized "rigidity" of the pattern. For $\varepsilon > 0$ the emerging structure will be approximately periodic of wavelength $\lambda_0 = 2\pi/q_0$. In an extended system of size $L \gg \xi_0 \gg \lambda_0$ (large aspect ratio), the coherence of the pattern will be maintained over distances proportional to ξ_0.

Pattern selection in confined geometries often can be understood in terms of the spatial modes of the base state that become linearly unstable and the geometry of the system. Extended systems, on the other hand, allow a richer nonlinear competition of linearly unstable modes, in part due to the restoration of some of the symmetries of the original system that had been broken by the boundaries. The distinction can be further quantified by using the coherence length of the pattern ξ_0, the linear size of the system L, and the characteristic wavelength λ_0. Directional solidification from the melt is a typical example of a system in the limit $\xi_0 \ll \lambda_0 \ll L$, fact that follows from the extremely flat neutral stability curve [1]. Experiments often show a narrow distribution of wavelengths in the stationary state, the prediction of which has proven elusive. Attempts at deriving amplitude equations valid near onset have not been successful precisely because of the condition $\xi_0 \ll \lambda_0$. A typical and widely studied example of the intermediate range concerns Rayleigh-Benard convection in large aspect ratio cells $\lambda_0 \precsim \xi_0 \ll L$. Although amplitude equations predict the existence of stable parallel rolls above onset, such a state is not generically observed in fluids of low or moderate Prandtl number except under carefully prepared initial conditions. Instead, a spatio-temporally chaotic state emerges that has been termed spiral defect chaos [2], [3]. The role that long wavelength modes (or mean flow) play in the de-

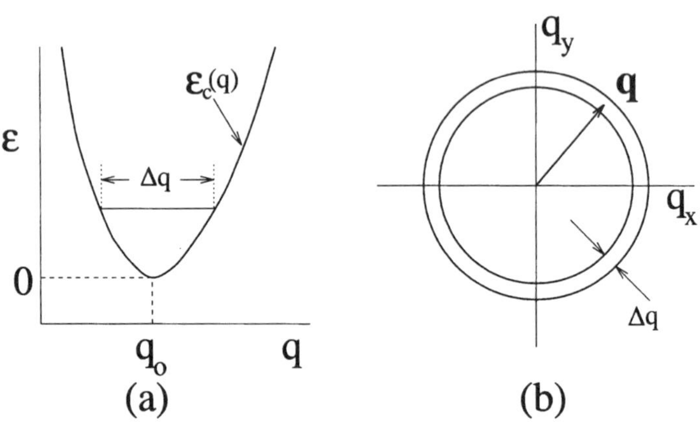

FIGURE 1. **(a)** Schematic representation of the neutral stability curve. **(b)** Degeneracy in wavevectors (for a two-dimensional system which is invariant under rotation) showing the band of unstable wavevectors.

velopment of such a state is still a matter of research [4]. Faraday waves, on the other hand, are a prototypical case of a system with a large coherence length $\xi_0 \gg \lambda_0$.

Another issue that may become important in an extended system is related to possible degeneracies of the unstable spectrum due to some symmetry of the system. An important example for our purposes concerns an effectively two-dimensional system invariant under rotation. In this case, there is a continuum of unstable wavevectors $\mathbf{q}_0$ that lie on the circle $\|\mathbf{q}_0\| = \mathbf{q}_0$ (FIG. 1b). One may anticipate the existence of regions within which the pattern is locally periodic with wavevectors of a particular orientation. The system will then be comprised of a large number of such regions with arbitrary orientations of their local wavenumber. In some cases, such a structure is known to coarsen and evolve towards a final stationary state given by a single orientation [5], [6].

The amplitude equation formalism has been widely used to describe the weakly nonlinear regime above threshold [7], [8]. For small ε, growth of the unstable mode is slow, and can be described in a time scale $T = \varepsilon t$. According to Eq. (2), the emerging pattern is periodic with wavelength λ_0, modulated over distances of the order of $\xi_0/\sqrt{\varepsilon}$ (or, equivalently, modulated in the scale $X = \varepsilon^{1/2}x$). Hence, the study of the nonlinear evolution of the base pattern in Eq. (1) can be recast into the analysis of its modulation which naturally evolves in the scale T and changes over distances of scale X. These issues can be easily illustrated in the case of the Swift-Hohenberg equation [8], [9], a model of Rayleigh-Benard convection near onset,

$$\frac{\partial \psi}{\partial t} = \varepsilon\psi - (q_0^2 + \partial_x^2)^2\psi - \psi^3, \tag{4}$$

where ψ is related to the velocity field at the mid-plane of the fluid layer and $\lambda_0 = 2\pi/q_0$ is the wavelength of the convective rolls. By defining $X = \varepsilon^{1/2}x$, $T = \varepsilon t$ and, $\psi(x, t) = \varepsilon^{1/2}\psi_0 + \varepsilon\psi_1 + \varepsilon^{3/2}\psi_2 + \ldots$, one obtains a (singular) perturbation expansion in ε. At order $\varepsilon^{1/2}$ one recovers the linear stability problem. The solution at this order is,

$$\psi_0 = A_0(X, T)\, e^{iq_0x} + A_0^*(X, T)\, e^{-iq_0x}, \tag{5}$$

where the amplitudes are (undetermined) functions of the slow variables. The equations at order ε are identically satisfied. At order $\varepsilon^{3/2}$, solvability condition appears of the form,

$$\frac{\partial A_0}{\partial T} = A_0 + 4q_0^2\partial_X^2 A_0 - 3|A_0|^2 A_0$$

or (6)

$$\partial_t A = \varepsilon A + \xi_0^2\partial_x^2 A - 3|A|^2 A,$$

with $A = \varepsilon^{1/2}A_0$. Equation (6) is known as the real Ginzburg-Landau equation, and lies at the heart of many pattern forming instabilities. Its form is, in fact, generic for stationary supercritical bifurcations [8], and hence has been studied in a wide variety of physical contexts. The equation describes the evolution of the amplitude of the unstable mode into the nonlinear regime, and, in particular, predicts saturation at a value $|A| = \sqrt{\varepsilon/3}$.

In the remainder of this paper, we will briefly summarize two systems in which a straightforward application of this formalism fails to capture their main qualitative features: spiral defect chaos in Rayleigh-Bénard convection, and the existence of quasi-periodic patterns in parametrically driven surface waves. In the former case, unstable modes of wavenumbers close to q_0 induce, and couple through, long wavenumber modes ($q \approx 0$) which are only weakly damped in fluids of small viscosity. Neglecting such modes leads to predictions of stable patterns above onset that are not generically observed experimentally in large aspect ratio systems. In the latter case, unstable modes of wavenumber q_0, that couple quadratically through modes of wavenumber $2q_0$, can in fact achieve resonance with these higher harmonics in a way that determines the selected pattern above onset. The resonant angle is a function of the base wavenumber q_0, which can be modified by varying, for example, the driving frequency in the experiment. When the resonant angle is almost zero, patterns of high symmetry (hexagonal, sixfold, eightfold, ...) are selected.

The transition from a parallel roll state to a chaotic pattern (which is referred to as spiral defect chaos) was observed recently in Rayleigh-Bénard convection, for large aspect ratio, at a moderate Rayleigh number and a low Prandtl number [2]. This is a classical configuration in which a fluid layer is heated from below, such that above a certain value of the temperature difference across the layer (the appropriate dimensionless group is the Rayleigh number), the quiescent and conductive state becomes linearly unstable to convective rolls. The chaotic state observed recently is comprised of a large number of rotating spirals which nucleate, interact, and annihilate, yielding a macroscopically disordered pattern. This state has been qualitatively modeled by using a generalized Swift-Hohenberg equation based on Eq. (4) [3]. Earlier theories based on amplitude equations predict straight rolls that are linearly stable in the region in which spiral chaos has been observed experimentally [10].

A minimal model which is sufficient to describe the qualitative features of this state is [11], [12]

$$\frac{\partial \psi(\mathbf{r}, t)}{\partial t} + \mathbf{u} \cdot \nabla \psi = [\varepsilon - (\nabla^2 + 1)^2]\psi - \psi^3, \tag{7}$$

$$\left[\frac{\partial}{\partial t} - Pr(\nabla^2 - c^2)\right]\nabla^2 \zeta = [\nabla(\nabla^2 \psi) \times \nabla \psi] \cdot \hat{e}_z, \tag{8}$$

where $\mathbf{r}$ is the position and $\mathbf{u}$ the velocity on the plane normal to the temperature gradient (along z), $\nabla^2 = \partial_x^2 + \partial_y^2$. Since the fluid is incompressible, the velocity $\mathbf{u}$ has been written in terms of the stream function (as, $\mathbf{u} = (\partial_y \zeta)\hat{e}_x - ((\partial_x \zeta)\hat{e}_y$. Equation (7) with $\mathbf{u} = 0$ reduces to the Swift-Hohenberg equation (4). Horizontal flow arises through vertical vorticity creation by roll curvature and amplitude modulations (right-hand side of Eq. (8)). Pr is the Prandtl number, and c^2 is an unknown constant which models viscous damping of the horizontal flow component by the top and bottom cell walls. Coupling to mean horizontal flow had been shown to play a key role, for example, in the onset of weak turbulence in Boussinesq fluids [13], [14]. The same coupling is sufficient to describe the chaotic state. We point out that the flow driven by roll curvature in (8) is of long wavelength character ("mean flow"), and is related to a uniform translation mode that is marginal for stress free boundary conditions at the top and bottom walls. For fluids of small Pr, this mode is only weakly damped through the formation of a small boundary layer adjacent to the walls.

2. PATTERNS AND QUASI-PATTERNS IN FARADAY WAVES

Parametrically driven surface waves (or Faraday waves) appear on the free surface of fluid layer when an imposed periodic vibration in the direction normal to the surface at rest is large enough to overcome the dissipative effect of fluid viscosity [8], [15]. Different wave patterns appear depending on the fluid properties and the driving amplitude or frequency. Experimental evidence indicates that the observed bifurcating wave pattern consists of parallel stripes (a single plane wave) at high viscous dissipation (a high viscosity fluid and/or low driving frequency) [16], [17]. For lower dissipation, patterns of square symmetry (combinations of two perpendicular plane waves) have been observed in the capillary regime (large frequencies) [18]. At lower frequencies (the mixed gravity-capillary regime), hexagonal, eightfold, and tenfold patterns have been observed [19], [20].

Progress in deriving suitable amplitude equations for Faraday waves has proved difficult. Since Faraday waves are almost a Hamiltonian system (weakly dissipative), most analyses are based on a Hamiltonian description for the ideal (inviscid) limit, and treat viscous or dissipative effects as a perturbation [21], [22]. The central question that arises in this case concerns the origin of any nonlinear saturation of the wave pattern since general symmetry considerations for Hamiltonian systems ("chiral symmetry") prohibit cubic terms in standing wave-amplitude equations [8], [23]. A recent analysis has explicitly addressed these issues in the limit of small fluid viscosity [24], [25]. A set of quasi-potential equations was derived by considering the rotational flow within a small viscous layer near the free surface, and assuming potential flow in the bulk. A standing wave-amplitude equation was obtained with cubic order terms given explicitly. Although this latter model predicted bifurcations to square and higher-symmetry patterns, it did rely on an uncontrolled approximation concerning nonlinear viscous terms. As a consequence, its region of validity is difficult to assess. In particular, it failed to yield the observed stripe patterns at intermediate and large viscosities. We recently overcame these difficulties, and systematically derived a standing wave-amplitude equation without any assumption other than those inherent in the multiple scale expansion [26]. As part of the derivation, we also obtained an analytical expression for the linear threshold of instability, which previously was known only numerically [27].

A. Governing Equations and Linear Stability

Consider a semi-infinite fluid layer, unbounded in the $x - y$ direction, extending to $z = -\infty$, and with a planar free surface at $z = 0$ when at rest. The fluid is assumed incompressible and Newtonian. Under periodic vibration of the layer in the direction normal to the surface at rest, the equation governing fluid motion (in the co-moving reference frame) is

$$\partial_t \mathbf{u} + (\mathbf{u} \cdot \nabla)\mathbf{u} = -\frac{1}{\rho} \nabla p + \upsilon \nabla^2 \mathbf{u} + g_z(t)\hat{\mathbf{e}}_z, \tag{9}$$

with $\mathbf{u} = (u, v, w)$ the velocity field, p the pressure, ρ and ν the density and kinematic viscosity of the fluid, respectively, and $g_z(t) = -g - f \cos\omega t$ the effective gravity. The base state is a quiescent fluid with a pressure distribution $p = \rho g_z(t)z$. Besides the null conditions at $z = -\infty$, there are four boundary conditions at the moving free surface [28]. Let $z = \zeta(x, y)$ be the position of the surface, $\hat{\mathbf{n}}$ the outward pointing unit normal and $\hat{\mathbf{t}}_1$ and $\hat{\mathbf{t}}_2$ two unit tangential vectors. The kinematic boundary condition is,

$$\partial_t \zeta + (\mathbf{u}(z = \zeta) \cdot \nabla_H)\zeta = w(z = \zeta),$$

with $\nabla_H = \hat{\mathbf{e}}_x \partial_x + \hat{\mathbf{e}}_y \partial_y$. Neglecting the effect of the air phase above the fluid, the tangential stress at the free surface is zero, $\hat{\mathbf{t}}_m \cdot \mathbf{T} \cdot \hat{\mathbf{n}}\big|_{z=\zeta} = 0$, $m = 1, 2$, with $\mathbf{T}$ the stress tensor of components

$$T_{ij} = [-p - \rho g_z(t) z]\delta_{ij} + \rho\nu(\partial_j u_i + \partial_i u_j).$$

The normal stress at the fluid surface is balanced by capillarity, $\hat{\mathbf{n}} \cdot \mathbf{T} \cdot \hat{\mathbf{n}}\big|_{z=\zeta} = 2H\sigma$ where σ is the surface tension and $2H$ is the mean curvature of the surface,

$$2H =$$

$$\{(\partial_{xx}\zeta)[1 + (\partial_y\zeta)^2] + (\partial_{yy}\zeta)[1 + (\partial_x\zeta)^2] - 2(\partial_x\zeta)(\partial_y\zeta)(\partial_{xy}\zeta)\}/[1 + (\partial_x\zeta)^2 + (\partial_y\zeta)^2]^{3/2}.$$

The linear stability of the fluid layer under vibration was first addressed in the inviscid limit [29], and later in the limit of low viscosity [30]. More recently, Kumar and Tuckerman [27] numerically computed the neutral stability curve for a fluid of arbitrary viscosity, and Müller *et al.* [31] have given an analytic low viscosity expansion. The dominant response of the parametrically driven system is subharmonic at a frequency $\omega/2$ [29]. To address the linear stability of the fluid surface, one considers the following solutions for the z component of the velocity field and surface displacement,

$$\begin{aligned} w_0 &= \cos(qx) \sum_{j=1,3,5,\ldots} e^{ji\omega t/2} w_0^j(z) A_j + \text{c.c.}, \\ \zeta_0 &= \cos(qx) \sum_{j=1,3,5,\ldots} e^{ji\omega t/2} A_j + \text{c.c.}, \end{aligned} \tag{10}$$

where the A_j are complex amplitudes, and we retain all the harmonics of the fundamental mode $e^{i\omega t/2}$. In addition to having two independent variables, this expression is more complicated than Eq. (1) because the bifurcating state is a wave of finite frequency (Hopf bifurcation) instead of the stationary bifurcation discussed in the introduction. The dispersion relation for *inviscid* surface waves is [32]

$$\omega_0^2 = g q_0 + \sigma q_0^3/\rho, \tag{11}$$

with $\omega_0 = \omega/2$ due to subharmonic resonance. In a fluid of low viscosity, q_{onset} is expected to be close to q_0. It is then convenient to define dimensionless variables by using $1/\omega_0$ as the time scale, and $1/q_0$ as the length scale. We also define a reduced wavenumber $\bar{q} = q/q_0$, a viscous damping coefficient $\gamma = 2\nu q_0^2/\omega_0$ the gravity wave $G = g q_0/\omega_0^2$ and capillary wave $\Sigma = \sigma q_0^3/\rho\omega_0^2$ contributions to the dispersion relation, and the dimensionless amplitude of the driving acceleration $\Delta = f q_0/4\omega_0^2$. Note that $G + \Sigma = 1$ from Eq. (11); $G = 1$ corresponds to a pure gravity wave while $G = 0$ to a pure capillary wave. For $\gamma \ll 1$, q_{onset} and Δ_{onset} can be given as a power series of the damping coefficient γ

$$\bar{q}_{\text{onset}} = 1 + \frac{1}{3-2G}\gamma^{3/2} + \frac{-7+2G}{(3-2G)^2}\gamma^2 + \ldots, \tag{12}$$

$$\Delta_{\text{onset}} = \gamma - \frac{1}{2}\gamma^{3/2} + \frac{11+2G}{8(3-2G)^2}\gamma^{5/2} + \ldots. \tag{13}$$

The first correction term is proportional to $\gamma^{3/2}$, and agrees with a low viscosity expansion of the linearized equations given in [31]. As a reference, we note that a similar linear analysis based on an inviscid formulation to which viscosity is added through a dissipation function, leads to the the damped Mathieu equation [30],

$$\partial_t^2 h_q(t) + \gamma \partial_t h_q(t) + \omega_0^2(1 + 2\Delta\cos 2\omega_0 t)h_q(t) = 0$$

with h_q the amplitude of the spatial mode with wavenumber q. This equation gives a threshold at $\gamma + 3\gamma^2/64 + O(\gamma^3)$. The first correction term is of a different order and has a different sign.

B. Standing Wave-Amplitude Equation and Pattern Selection

In this section, we summarize the results of a multiple scale approach to derive standing wave-amplitude equations valid near threshold [26]. For a driving amplitude f above threshold f_0, $\varepsilon = (f - f_0)/f_0$ and the flow is expanded as

$$\mathbf{u} = \varepsilon^{1/2}\mathbf{u_0} + \varepsilon\mathbf{u_1} + \varepsilon^{3/2}\mathbf{u_2} + \ldots,$$

and similarly p and ζ. Near threshold, i.e., for $\varepsilon \ll 1$, $T = \varepsilon t$, $\partial_t \to \partial_t + \varepsilon\partial_T$. Spatial slow scales are not included because only regular patterns are considered here. At order $\varepsilon^{1/2}$ one recovers the linear problem discussed in the previous section. Since we are interested in standing wave patterns of different symmetries, the solution at this order is written as a linear combination of waves with wavevectors q_m of magnitude q_{onset} but along different directions on the $x - y$ plane:

$$w_0 = \sum_m \cos(\mathbf{q}_m \cdot \mathbf{x})B_m(T) \sum_{j=1,3,5,\ldots} e^{ji\omega t/2} w_0^j(z)e_j + \text{c.c.},$$

$$\zeta_0 = \sum_m \cos(\mathbf{q}_m \cdot \mathbf{x}) \sum_{j=1,3,5,\ldots} e^{ji\omega t/2} e_j + \text{c.c.},$$

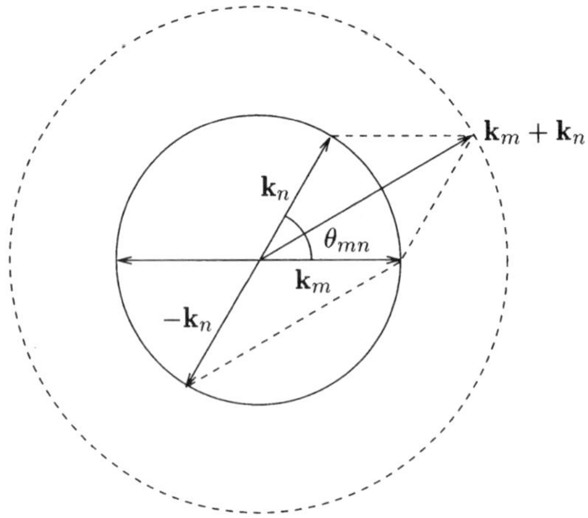

FIGURE 2. Schematic representation of a triad resonant interaction: two linearly unstable modes $\mathbf{k}_m$ and $\mathbf{k}_n$ interact to produce a linearly stable mode. This mode interacts with $-\mathbf{k}_n$ leading to a resonance with $\mathbf{k}_m$.

where $B_m(T)$ are *real* wave-amplitudes, functions only of the slow time scale T, and the e_j are the same as the A_j found in Eq. (10).

At order ε the equation of motion is fairly complicated and will not be reproduced here. At order $\varepsilon^{3/2}$ a solvability conditions appears that leads to a standing wave-amplitude equation for a particular unstable mode B_1,

$$\frac{dB_1}{dT} = \alpha B_1 - g_0 B_1^3 - \sum_{m \neq 1} g(\theta_{m1}) B_m^2 B_1 , \tag{13}$$

with θ_{m1} the angle between $\mathbf{q}_m$ and $\mathbf{q}_1$. This equation is of the same form as Eq. (6), but with a cubic coefficient that couples standing waves of different orientations. The linear coefficient α is the linear growing or decaying rate of this wave, and can be obtained from the linear analysis. The coefficient $g(\theta)$ describes the nonlinear interaction between different linearly unstable modes, and provides for the saturation of the wave amplitude. A distinctive feature of this problem is the existence of triad resonant interactions. FIGURE 2 shows two linearly unstable modes with wavevectors $\mathbf{k}_m$ and $\mathbf{k}_n$ ($|\mathbf{k}_m| = |\mathbf{k}_n| = k_{\text{onset}}$). They interact to produce a wave at $\mathbf{k}_m + \mathbf{k}_n$ with an amplitude proportional to $B_m B_n$. This mode corresponds to a first-order solution. Now $\mathbf{k}_m + \mathbf{k}_n$ couples back to the original wave at $-\mathbf{k}_n$, to give a contribution $B_n^2 B_m$ to dB_m/dT. Since the mode $\mathbf{k}_m + \mathbf{k}_n$ is damped (only waves with wavenumbers near k_{onset} are unstable), this is a dissipative term and contributes to nonlinear saturation of the wave.

Triad resonance was already encountered by Milner [21] as a divergence in the coefficients of the cubic terms of the amplitude equation at a particular value of the angle θ_{ml}. Edwards and Fauve [16] later noted that for a weakly damped system, triad resonance could be important for pattern selection, issue that was further studied in [25]. Triad resonance occurs when the frequency of the mode $\mathbf{k}_m + \mathbf{k}_n$ equals the driving frequency (the modes $\mathbf{k}_m$ and $\mathbf{k}_n$ oscillate at half the driving frequency). Energy is now directly transferred into this mode which can have a very large amplitude at low damping. Since $\mathbf{k}_m + \mathbf{k}_n$ couples back to $-\mathbf{k}_n$, and thus provides a dissipation channel for the mode $\mathbf{k}_n$, dissipation is enhanced by triad resonance and results in a large value of $g(\theta_{mn})$ in the vicinity of the resonant angle. The resonant angle can be estimated from the inviscid dispersion relation (11), written in dimensionless form,

$$\bar{\omega}^2 = G\bar{q} + \Sigma\bar{q}^3, \tag{14}$$

with $\bar{q} = 1$, and $\bar{\omega}^2 = G + \Sigma = 1$ for the linearly unstable mode. At resonance, we have $\bar{\omega} = 2$, and the resonant wavenumber $\bar{q}_r = |\mathbf{q}_m + \mathbf{q}_n|$ satisfies $\bar{q}_r(G + \Sigma\bar{q}_r^2) = 4$. If θ_r is the resonant angle between $\mathbf{q}_m$ and $\mathbf{q}_n$, $\bar{q}_r = \sqrt{2(1+\cos\theta_r)}$ the resonance condition becomes

$$\sqrt{2(1+\cos\theta_r)}\,[G + 2(1+\cos\theta_r)\Sigma] = 4. \tag{15}$$

Because $G + \Sigma = 1$, this condition can only be satisfied when $\Sigma > 1/3$. More generally, three wave-resonant interactions are permitted when the dispersion relation of the wave satisfies $\partial^2\omega/\partial q^2 \geq 0$, as is the case in the capillary regime. For gravity waves $\partial^2\omega/\partial q < 0$, and only four wave-resonant interactions are possible. For finite damping, the resonance condition is modified. However, triad resonance is expected to be significant only at low damping because of the damped nature of the stable wave. At small γ, the nonlinear coefficient grows near resonance and peaks at the resonant angle. The value of the peak is seen to decrease with increasing γ.

Since the standing wave-amplitude equation (13) can be written in gradient form, the selected pattern near threshold immediately follows [8]). Equation (13) is equivalent to

$$\frac{dB_n}{dT} = -\frac{\delta\mathcal{F}}{\delta B_n},$$

with the Lyapunov function $\mathcal{F}$ given by

$$\mathcal{F} = -\frac{1}{2}\alpha\sum_m B_m^2 + \frac{1}{4}\sum_m\sum_n g(\theta_{mn})B_m^2B_n^2,$$

with $g_0 = g(\theta_{nn})$ which equals half the value of $g(\theta \to 0)$. The amplitude equation then implies that

$$\frac{d\mathcal{F}}{dT} = \sum_n \frac{\delta\mathcal{F}}{\delta B_n}\frac{dB_n}{dT} = -\sum_n\left(\frac{dB_n}{dT}\right)^2 \leq 0,$$

so that the preferred pattern can be determined by minimization of $\mathcal{F}$. The experimentally observed regular patterns above onset consist of N standing waves, with uniform amplitudes and wavevectors $\mathbf{q}_n$, $m = 1 \ldots N$. $N = 1$ corresponds to a single standing wave (a pattern of parallel stripes), $N = 2$ to a pattern of square symmetry, $N = 3$ of hexagonal symmetry, etc. For these regular patterns, the steady states wave amplitudes are

$$B_n^2 = \frac{\alpha}{g_0 + \Sigma_{m \neq n} g(\theta_{mn})}, \quad n = 1 \ldots N. \tag{15}$$

The value of the Lyapunov function as a function of N then becomes

$$\mathcal{F}(N) = -\frac{\alpha^2}{4} \frac{N}{g_0 + \Sigma_{m=2}^{N} g(\theta_{m1})}. \tag{16}$$

For pure gravity waves ($\Sigma = 0$), the $N = 1$ state has the lowest value of the Lyapunov function and hence will be the selected pattern. This results differs from the predictions of the low damping calculation of in [25]. At low frequencies, the system effectively crosses over to the large damping region regardless of its (finite) viscosity. On the other hand, for pure capillary waves ($\Sigma = 1$) the preferred pattern is $N = 2$ at low damping, and $N = 1$ at high damping. Interesting behavior is observed in the vicinity of $\Sigma = 1/3$ (mixed gravity capillary waves) where the triad resonance angle approaches zero. Hexagonal and higher symmetry quasi patterns are selected with decreasing γ. The low damping results in this region are in agreement with earlier work by Zhang and Viñals [25], although the latter could not account for the transition to $N = 1$ as γ is increased.

We finally compare our predictions (based on Eq. (16)), and two recent sets of experiments that addressed the large aspect ratio behavior [19], [20]. The only input parameters in our calculations are the fluid properties (density, surface tension, and viscosity), and the frequency of the driving acceleration. All these parameters are known fairly precisely in the experimental work, thus allowing a quantitative comparison between theory and experiments. Pattern selection in the low viscosity range has been studied in [20]. They have developed a cell of exceptionally large aspect ratio, and depth which is much larger than the wavelength. When the driving frequency is decreased from 45 Hz, a transition from a $N = 2$ square pattern to an $N = 3$ hexagonal pattern was observed at approximately 35 Hz, and to a quasi-periodic $N = 4$ eightfold pattern at approximately 29 Hz. The prediction for these transitions based on Eq. (16) are 35.4 Hz and 28.7 Hz, respectively. These results are also in good agreement with the earlier weak damping calculation that predicted the same transitions at frequencies of 32.8 Hz and 27.9 Hz, respectively [24].

Another set of recent experiments involving fluids of different viscosity has been described in [19]. Although the depth of the fluid layer (0.3 cm) is smaller than the wavelength of the waves (1–3 cm), the comparison is still illuminating. FIGURE 3 shows the symmetry of the predicted patterns as a function of the viscosity of the fluid and of the driving frequency (with $\rho = 0.95$ g/cm^3 and $\sigma = 20.6$ dyn/cm), and the experimentally observed patterns. They are stripes at high viscosity, hexagons at low viscosity and frequency, and squares at low viscosity and high frequency. Two significant discrepancies concern the experimental observation of a hexagonal pattern at $\nu = 1$ cm^2/s and low frequency, and also at $\nu = 0.04$ cm^2/s and $f = 27$ Hz. It

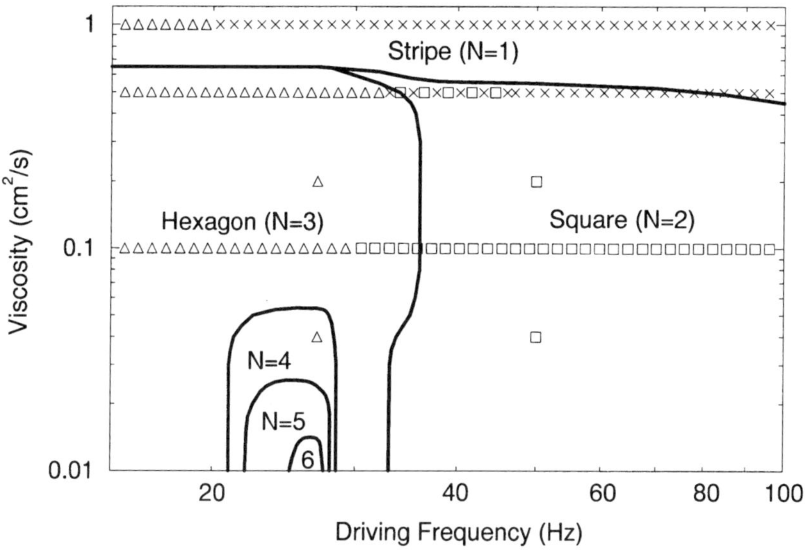

FIGURE 3. Selected patterns as a function of fluid viscosity and driving frequency. The symbols are the experimental results of [19]. Key: × — stripe patterns; □ — square patterns; △ — hexagonal patterns. Alternating × and □ indicate regions in which stationary mixed stripe and square patterns were observed.

is possible that the shallowness of the fluid layer can account for these differences, especially in view of the fact that, as noted above, the experiments in [20] did probe this latter region in a deep fluid layer, and their results do agree with our predictions.

The transition from square to stripe patterns remains in the capillary wave limit of $\Sigma = 1$ (high frequency limit in the experiment). However, the figure for Σ_0 indicates that stripe patterns are always preferred in the pure-gravity wave limit (low frequency in the experiment, not shown in FIG. 3). Furthermore, all the high symmetry patterns (with $N \geq 3$) are observed in the vicinity of $\Sigma = 1/3$, the point at which the triad resonant angle approaches zero, and for low damping where the resonance is more pronounced. Finally, the fact that portions of the boundaries separating regions of different symmetry appear almost as straight lines in FIG. 3 is due to the log–log scale used. In addition, the transition line delimiting the region of stripe patterns is almost independent of frequency only because of the limited frequency range probed in the experiments and displayed in the figure. And, the line separating square and hexagonal patterns is almost independent of viscosity because it depends mainly on whether the waves are capillarity or gravity dominated, a fact that is largely dependent on driving frequency and not on viscosity.

ACKNOWLEDGMENTS

I am indebted to a number of collaborators for the research presented. Work on Rayleigh-Benard convection has been done in collaboration with Hao-wen Xi and

Jim Gunton, and work on Faraday waves in collaboration with Wenbin Zhang and Peilong Chen.

REFERENCES

1. MULLINS, W.W. & R.F. SEKERKA. 1964. Stability of a planar interface during solidification of a dilute binary alloy. J. Appl. Phys. **35:** 444–451.
2. MORRIS, S., E. BODENSCHATZ, D. CANNELL & G. AHLERS. 1993. Spiral Defect Chaos in Large Aspect Ratio Rayleigh-Benard Convection. Phys. Rev. Lett. **71:** 2026–2029.
3. XI, H., J. GUNTON & J. VINALS. 1993. Spiral defect chaos in a model of RayleighBenard convection. Phys. Rev. Lett. **71:** 2030–2033.
4. CROSS, M. & Y. TU. 1995. Defect dynamics for spiral chaos in Rayleigh-Benard convection. Phys. Rev. Lett. **75:** 834–837.
5. ELDER, K., J. VIRALS & AND M. GRANT. 1992. Ordering dynamics in the twodimensional stochastic Swift-Hohenberg equation. Phys. Rev. Lett. **68:** 3024–3027.
6. CROSS, M. & D. MEIRON. 1995. Domain coarsening in systems far from equilibrium. Phys. Rev. Lett. **75:** 2152–2155.
7. NEWELL, A. & J. WHITEHEAD. 1969. Finite bandwidth, finite amplitude convection. J. Fluid Mech. **38:** 279–303.
8. CROSS, M. & P. HOHENBERG. 1993. Pattern formation outside of equilibrium. Rev. Mod. Phys. **65:** 851–1112.
9. SWIFT, J. & P. HOHENBERG. 1997. Hydrodynamic fluctuations at the convective instability. Phys. Rev. A. **15:** 319–328.
10. BUSSE, F. 1978. Non-linear propoerties of thermal convection. Rep. Prog. Phys. **41:** 1921–1967.
11. CROSS, M. 1980. Derivation of the amplitude equation at the Rayleigh-Benard instability. Phys. Fluids. **23:** 1727–1731.
12. GREENSIDE, H., W. COUGHRAN, JR. & N. SCHRYER. 1982. Nonlinear Pattern Formation near the Onset of Rayleigh-Benard Convection. Phys. Rev. Lett. **49:** 726–729.
13. SIGGIA, E. & A. ZIPPELIUS. 1981. Pattern Selection in Rayleigh-Benard Convection near Threshold. Phys. Rev. Lett. **47:** 835–838.
14. MANNEVILLE, P. 1983. A two-dimensional model for three-dimensional convective patterns in wide containers. J. Physique 44: 759–765.
15. MILES, J. & D. HENDERSON. 1990. Paranetrically forced surface waves. Annu. Rev. Fluid Mech. **22:** 143–165.
16. EDWARDS, S. & S. FAUVE. 1994. Patterns and quasi-patterns in the Faraday experiment. J. Fluid Mech. **278:** 123–148.
17. DAUDET, L., V. EGO, S. MANNEVILLE & J. BECHHOEFER. 1995. Secondary instabilities of surface waves on viscous fluids in the Faraday instability. Europhys. Lett. **32:** 313–318.
18. TUFILLARO, N., R. RAMSHANKAR & J. GOLLUB. 1989. Order-Disorder Transition in Capillary Ripples. Phys. Rev. Lett. **62:** 422–425.
19. KUDROLLI, A. & J. GOLLUB. 1996. Patterns and Spatiotemporal Chaos in Parametrically Forced Surface Waves: A Systematic Survey at Large Aspect Ratio. Physica D. **97:** 133–154.
20. BINKS, D. & W. VAN DE WATER. 1997. Nonlinear pattern formation of Faraday waves. Phys. Rev. Lett. **78:** 4043–4046.
21. MILNER, S. 1991. Square patterns and secondary instabilities in driven capillary waves. J. Fluid Mech. **225:** 81–100.
22. MILES, J. 1993. On Faraday waves. J. Fluid Mech. **248:** 671–683.
23. COULLET, P., T. FRISCH & G. SONNINO. 1994. Dipersion-induced patterns. Phys. 16 Rev. E. **49:** 2087–2090.

24. ZHANG, W. & J. VIRALS. 1996. Square patterns and quasi-patters in Faraday waves in low viscosity fluids. Phys. Rev. E. **53:** R4283–R4286.
25. ZHANG, W. & J. VINALS 1997. Pattern formation in weakly damped Faraday waves. J. Fluid Mech. **336:** 301–330.
26. CHEN, P. & J. VINALS. 1997. Pattern Relection in Faraday Waves. Phys. Lett. **79:** 2670–2673.
27. KUMAR, K. & L. TUCKERMAN. 1994. Parametric instability of the interface between two fluids. J. Fluid. Mech. **279:** 49–68.
28. LAMB, H. 1932. Hydrodynamics. Cambridge University Press, Cambridge.
29. BENJAMIN, T. & F. URSELL. 1954. The stability of the plane surface of a liquid in vertical periodic motion. Proc. R. Soc. London Ser. A. **225:** 505–515.
30. LANDAU, L. & E. LIFSHITZ. 1976. Mechanics. Pergamon, New York.
31. MÜLLER, H. *et al.* 1997. Analytic Stability Theory for Faraday Waves and the Observation of the Harmonic Surface Response. Phys. Rev. Lett. **78:** 2357–2360.
32. LANDAU, L. & E. LIFSHITZ. 1959. Fluid Mechanics. Pergamon. New York.

Coupled Oscillators for Fun and Profit

KURT WIESENFELD

School of Physics, Georgia Institute of Technology, Atlanta, Georgia 30332

ABSTRACT: A prime example of long-range cooperative behavior is when populations of nonlinear oscillators spontaneously synchronize. This phenomenon crops up in many places in biology and applied physics. Some twenty years ago, Yoshiki Kuramoto introduced a simple tractable model which captures key elements of the phenomenon. It also provides new answers to old questions about certain superconducting circuits having applications for fast electronics.

MUTUAL SYNCHRONIZATION

More and more, researchers who study nonlinear dynamics are turning their attention toward so-called complex systems, i.e., systems which involve many interacting degrees of freedom. One area where there has been substantial progress is populations of coupled oscillators. An especially interesting phenomenon is mutual entrainment, wherein elements with different natural frequencies spontaneously synchronize to oscillate at some compromise frequency.

The simplest version of mutual synchronization involves just two oscillators. Perhaps the earliest documented study of the effect was recorded by Christien Huygens in 1665 in a letter to his father [1], [2] describing the correlated motion of two pendulum clocks. The essence of the phenomenon is this: if ω_1 and ω_2 are the two natural frequencies (i.e., when there is no coupling) and Ω_1 and Ω_2 are the dressed frequencies (i.e., the measured frequencies in the coupled case), then for strong enough coupling $\Omega_1 = \Omega_2$.

The situation becomes more interesting when there are many interacting oscillators with a distribution of natural frequencies. This situation arises in applied physics, for example, in studies of coupled lasers and of arrays of nonlinear electrical oscillators. In these examples the practical motivation is to boost the total output by using many elements instead of one. To achieve desirable levels of coherence it is important that the elements lock to the same frequency despite the inevitable variation in natural frequencies. Nonlinearity provides the essential a mechanism for such mutual synchronization.

FIGURE 1 shows the results of numerical simulations of a superconducting circuit consisting of 100 oscillators known as Josephson junctions. The oscillators all have different natural frequencies. When coupled, the oscillators show significant synchronization, in this case about half of the junctions are observed to oscillate at exactly the same frequency.

Another intriguing example can be found in the notes of Norbert Wiener [2], [3]. FIGURE 2 qualitatively reproduces his sketch of the electromagnetic radiation near the head of a relaxed human subject. Of the particular lineshape — a narrow line arising from a dip in a broader background — Wiener wrote [4], "Consider the possibility that the brain contains a number of oscillators of frequencies nearly 10 per

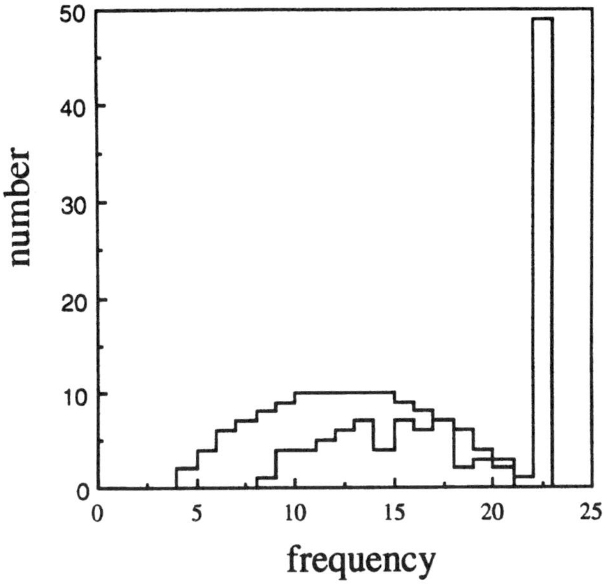

FIGURE 1. Histograms of natural and dressed frequencies for an array of 100 Josephson junctions, determined from numerical simulations of the nonlinear circuit equations.

second, and that … the frequencies are likely to be pulled together into one or more little clumps" due to their nonlinear interactions.

Other examples of mutual entrainment arise in biology [2], [5]. Take the behavior of a healthy human heart: the individual pacemaker cells have somewhat different natural frequencies but (thankfully) they mutually synchronize to beat in lock step. Apparently, mutual synchronization touches both the heart and mind of man.

THE KURAMOTO MODEL

It is not obvious that there is a tractable mathematical description of this behavior: it involves a large population of interacting nonidentical nonlinear oscillators, and we know that solving the equations of even a single nonlinear oscillator can be very difficult. Happily, there is a particular model which is both dynamically interesting and amenable to analysis [6]:

$$\dot{\theta}_k = \omega_k + \frac{K}{N}\sum_{j=1}^{N} \sin(\theta_j - \theta_k), \tag{1}$$

where $k = 1, \ldots, N$, and the natural frequencies $\{\omega_k\}$ are chosen from some given distribution. The parameter K sets the coupling strength: for $K = 0$ each (angular) variable θ_k rotates at its natural frequency ω_k. Two features of note are (i) every os-

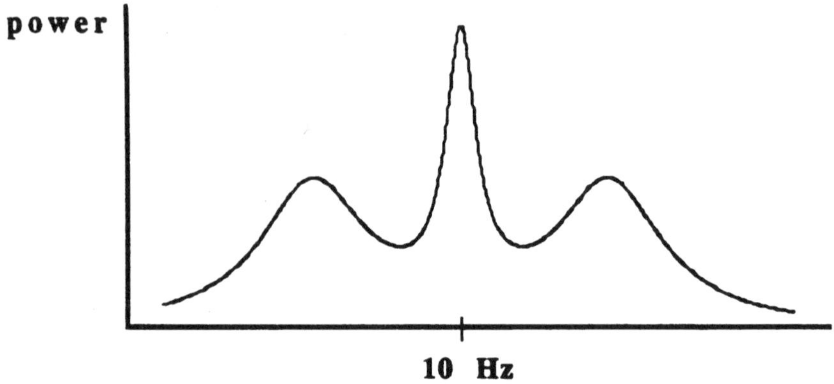

FIGURE 2. Wiener's sketch of the electromagnetic output as a function of frequency from a human subject (after [2]).

cillator is coupled to every other, and (ii) each oscillator is described by a single phase variable. The first feature may seem unphysical, but (as we'll see) there are cases where such global coupling arises naturally. The second feature is reasonable for weakly interacting oscillators.

Kuramoto intended this model to describe (if only crudely) a broad class of oscillator systems. The general meaning of the phase variable can be understood as follows. Consider first the uncoupled limit. Each oscillator is attracted to a limit cycle in the single-oscillator phase space, and the position along the limit cycle may be labelled by a periodic variable ϕ. This angle can be reparameterized so that the new angle θ advances uniformly. For weak coupling, it is reasonable to expect that the individual limit cycles do not substantially deform, and that the relevant dynamics takes place in the reduced N-dimensional phase space spanned by the N variables $\{\theta_k\}$.

The Kuramoto model displays the spontaneous cooperative behavior often seen in coupled oscillator systems. It is convenient to introduce a global measure of the coherence of the array

$$r\exp(i\psi) = (1/N)\sum_k \exp(i\theta_k)\,. \tag{2}$$

If you picture each of the phases θ_k as running around the unit circle, this is simply the centroid of the system. The degree of coherence of the N-oscillator system is reflected in r: if the θ_k are all equal then $r = 1$; if the θ_k are spread out uniformly then $r = 0$.

FIGURE 3 depicts the behavior of this model for large N. After a relatively brief transient time, r fluctuates about a constant value r_∞ while ψ rotates at a constant rate. Certain aspects regarding the fluctuations are not as yet understood, but they diminish with growing N, and for large enough N the asymptotic value r_∞ is apparently the same regardless of initial conditions, though it depends on the coupling

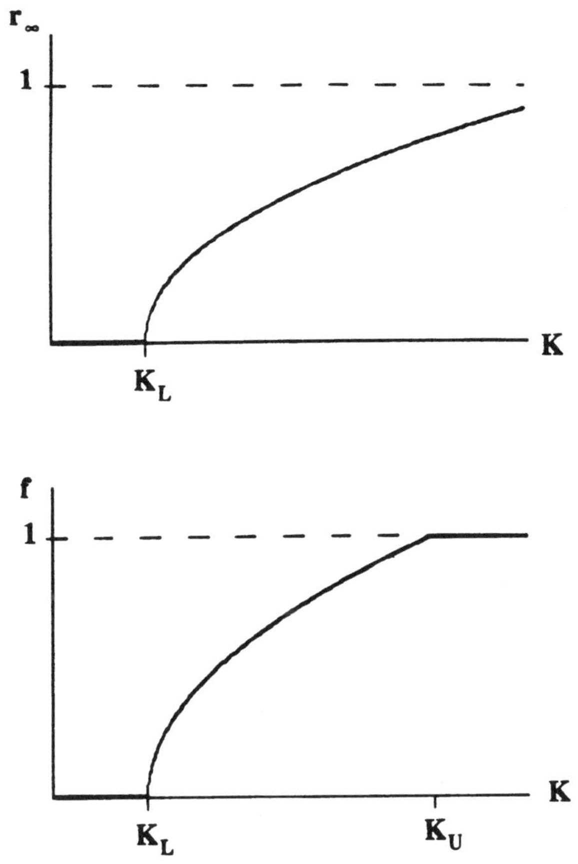

FIGURE 3. Typical large-N behavior of the Kuramoto model as a function of coupling strength K. **Top** — synchronization parameter r_∞; **bottom** — fraction f of frequency-locked oscillators.

constant K. There is a threshold value $K = K_L$ above which r_∞ is nonzero. Meanwhile, the fraction of oscillators which precisely frequency-lock behaves in a similar way except that there is a second transition at $K = K_U$ above which all N oscillators are synchronized.

The large-N behavior of Eq. (1) can be deduced using a self-consistency approach. Strogatz provides a crystal clear description of the method in [2]; further improvements (including stability analysis) are developed in [7]. The basic idea becomes evident after using Eq. (2) to rewrite Eq. (1) this way:

$$\dot{\theta}_k = \omega_k + Kr\sin(\psi - \theta_k). \tag{3}$$

Next we demand that both r and $d\psi/dt$ are constant — as observed in the numerical simulations — and solve Eq. (3) separately for each oscillator. Transforming to a frame rotating at frequency Ω, Eq.(3) becomes

$$\dot{\xi}_k = \omega_k - \Omega - Kr\sin\xi_k, \tag{4}$$

where $\xi_k = \theta_k - \Omega t$. We see that some of the oscillators are exactly entrained at the frequency Ω, since if $|\omega_k - \Omega| < Kr$ then according to Eq. (4), $\xi_k \to 0$ as time passes. Those oscillators, whose natural frequency lies outside this band, run at frequencies Ω_k different than Ω, in fact one easily shows that

$$\Omega_k = \Omega \pm \sqrt{(\omega_k - \Omega)^2 - (Kr)^2},$$

where the plus (minus) sign corresponding to the case where the natural frequency ω_k is greater (less) than the locking frequency Ω.

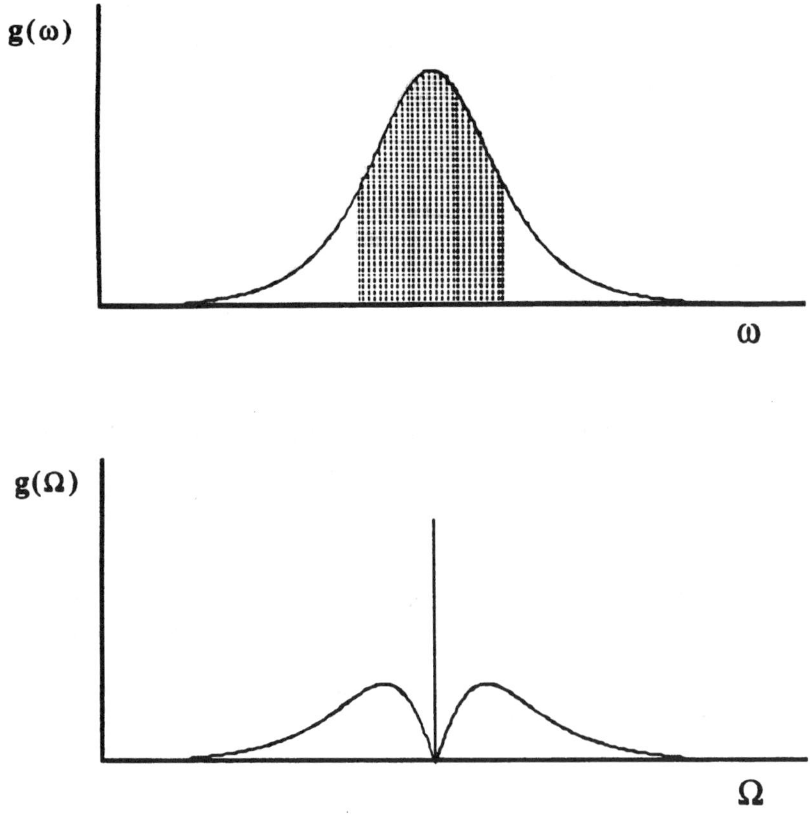

FIGURE 4. Example of the natural (**top**) and dressed (**bottom**) frequency distributions for the Kuramoto model in the range $K_L < K < K_U$.

The final step is to determine r and Ω by demanding that Eq. (2) is self-consistently satisfied. One finds that $r = 0$ is the only self-consistent solution for K less than a threshold value K_L. A typical situation for $K > K_L$ is depicted in FIG. 4 for a symmetric distribution: all oscillators in a band of frequencies (indicated by the shaded region of $g(\omega)$) synchronize at the mean frequency.

CONNECTION TO JOSEPHSON JUNCTION ARRAYS

The Kuramoto model is a useful paradigm: it shows many of the features observed in various types of oscillator arrays, though naturally different systems will show variants of the basic behavior. For the most part, Eq. (1) and its variants have been viewed as "toy models" which play a role akin to Ising spin models in statistical mechanics. However, the Kuramoto model is a quantitatively accurate description of a particular type of superconducting circuit, namely a current biased series array of Josephson junctions shunted by an external load.

Josephson arrays have been intensively studied because of their potential applications in very high frequency electronics. Ironically, this connection was recognized only recently [8], even though Josephson arrays have been around even longer than the Kuramoto model. The tractability of the Kuramoto model gives us a powerful tool to address an important practical question: how identical must the individual elements be in order to achieve full frequency locking?

Written in dimensionless form, the circuit equations for the Josephson array are

$$\dot{\phi}_k + I_k \sin\phi_k + \dot{Q} = I, \tag{6a}$$

$$L\ddot{Q} + R\dot{Q} + Q/C = \sum \dot{\phi}_j, \tag{6b}$$

where $k = 1, \ldots, N$ and the sum runs over all N values of j. The oscillators are different by virtue of the parameters I_k which represent the critical current of each superconducting junction. This system has some obvious similarities with the Kuramoto model: each oscillator is described by a single phase variable ϕ_k, and the coupling is global. Unlike the Kuramoto model, the coupling is mediated by another dynamical variable Q (representing the charge on the load capacitor): each oscillator is coupled to Q (Eq. (6a)), while Q is coupled to the sum of all N junction phases (Eq. (6b)). In fact, in the limit of weak coupling (small $\dot{Q}$) the circuit equations can be reduced using an averaging technique [9] to

$$\dot{\phi}_k = \omega_k + \frac{K}{N}\sum_{j=1}^{N} \sin(\theta_j - \theta_k + \alpha), \tag{7}$$

where the parameters K, α, and $\{\omega_k\}$ can be explicitly related to the original circuit parameters. The variables θ_k are reparameterized versions of ϕ_k as described earlier. This differs from Eq. (1) only by the presence of the parameter α, but this is only a minor nuisance and the system can be analyzed in the same way [10]. Thus, given the circuit parameters (including the distribution of junction crictical currents I_k), we

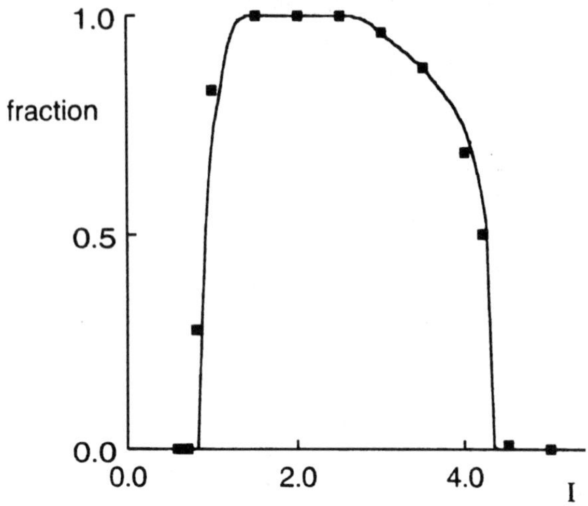

FIGURE 5. Comparison between the Kuramoto predictions (solid line) and numerical simulations of the full nonlinear circuit equations (spots) for the Josephson array.

can calculate the corresponding Kuramoto parameters, and determine the fraction of synchronized oscillators.

Are present fabrication techniques good enough to expect complete synchronization of this type of Josephson array? FIGURE 5 shows a comparison of the Kuramoto prediction (solid line) and the result of direct numerical integration of Eqs. (6) (spots) for an array of 100 junctions. The fraction of frequency-locked oscillators is plotted as a function of the bias current I. The parameters chosen are typical of real Josephson arrays; the agreement is excellent. (Note that while I is a natural control parameter, its relation with the Kuramoto coupling strength $K\cos a$ is nonmonotonic: decreasing I from a large value corresponds to first increasing and then decreasing the effective coupling, so that the full synchronization occurs over an interval of I.) One finds [8] that complete synchronization is achieved for variations in I_k less than about 1%, which as it happens is just about the limit of existing fabrication techniques [11]. FIGURE 1 corresponds to this same set of simulations where about half of the junctions are frequency locked.

REFERENCES

1. NIJHOFF, M., Ed. 1893. Oeuvres Completes des Christian Huygens. Societé Hollandaise des Sciences.The Hague. The Netherlands. **5:** 183–187.
2. STROGATZ, S.H. 1994. Norbert Wiener's Brain Waves. *In* Lecture Notes in Biomathematics. S. Levin, Ed. Vol. 100. Springer. Berlin.
3. WIENER, N. 1958. Nonlinear Problems in Random Theory. MIT Press. Cambridge, MA.
4. WIENER, N. 1961. Cybernetics. 2nd edit. 199. MIT Press. Cambridge, MA.
5. STROGATZ, S.H. & I. STEWART. 1993. Coupled oscillators and biological synchronization. Scientific American. **269:** (6) 102–109.

6. KURAMOTO, Y. 1975. Self-entrainment of a population of coupled nonlinear. *In* Lecture Notes in Physics. Vol. 39: 420–422. Springer. New York.
7. STROGATZ, S.H. & R.E. MIROLLO. 1991. Stability of incoherence in a population of coupled oscillators. J. Stat. Phys. **63:** 613–635.
8. WIESENFELD, K., P. COLET & S.H. STROGATZ. 1996. Synchronization transitions in a disordered Josephson series array. Phys. Rev. Lett. **76:** 404–407.
9. WIESENFELD, K. & JW. SWIFT. 1995. Averaged equations for Josephson junction series arrays. Phys. Rev. E. **51:** 1020–1025.
10. SAKAGUCHI, H. & Y. KURAMOTO. 1986. A soluble active rotator model showing phase transitions via mutual entrainment. Prog. Theor. Phys. 76: 576–581.
11. P.A.A. BOOI & S.P. BENZ. 1994. Emission linewidth measurements of two-dimensional array Josephson oscillators. Appl. Phys. Lett. **64:** 2163–2165.

Index of Contributors